NUCLEAR FACILITIES IN THE UNITED STATES

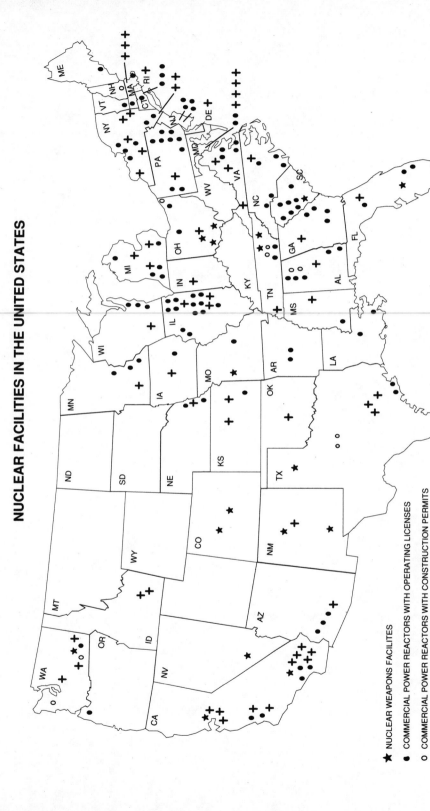

★ NUCLEAR WEAPONS FACILITES

● COMMERCIAL POWER REACTORS WITH OPERATING LICENSES

○ COMMERCIAL POWER REACTORS WITH CONSTRUCTION PERMITS

+ RESEARCH REACTORS

DERIVED FROM: NUCLEAR INFORMATION AND RESOURCE SERVICE, U.S. DEPARTMENT OF ENERGY, AND WAR RESISTERS LEAGUE

DEADLY DECEIT

DEADLY DECEIT

LOW-LEVEL RADIATION
HIGH-LEVEL COVER-UP

JAY M. GOULD AND
BENJAMIN A. GOLDMAN

WITH KATE MILLPOINTER

FOUR WALLS EIGHT WINDOWS

NEW YORK

Published by:
Four Walls Eight Windows
PO Box 548
Village Station
New York, N.Y. 10014

First edition.
First printing April 1990.

Much of Chapter Three originally appeared
in somewhat different form in *In These Times.*

Library of Congress Cataloging-in-Publication Data:

1. Nuclear power plants—Health aspects.   2. Nuclear power plants—
Environmental aspects.   I. Goldman, Benjamin A.   II. Title.
RA569.G68   1990   89-25688
363.17'99—dc20   CIP

ISBN: 0-941423-35-2

Text designed by Cindy LaBreacht

Printed in the U.S.A.

TABLE OF CONTENTS

ACKNOWLEDGMENTS

This book rests heavily on the insights and guidance of Dr. Ernest J. Sternglass, Professor Emeritus of Radiology at the University of Pittsburgh Medical School, Professor Jens Scheer of the University of Bremen, and of the late Dr. Carl Johnson, of the Colorado Department of Health. Dr. Johnson was the first epidemiologist to establish the link between bomb-test radiation and excess cancer rates in areas of Utah downwind of the Nevada Test Site. We also benefited greatly from critical comments from Dr. Donald B. Louria, Chairman of the Department of Preventive Medicine and Community Health of the New Jersey Medical School, and from Dr. Marvin Lavenhar, Director of the Biostatistics and Epidemiology Division of the New Jersey Medical School.

The book is the product of a collective effort on the part of consultants to, and the analysts and staff of, Public Data Access, Inc., including Ayse Cahn, Wendy Chiang, Bill McDonnell, Joseph Mangano, Chinnah Mithrasekaran, David Tanzer, Ken E. Tanzer, Ken J. Tanzer, Eben Weitzman, and Ellen Davidson who contributed their talents to the research and development of the databases that provided the radiation and mortality information used throughout the book.

Deadly Deceit benefited greatly from the journalistic skills of Kate Millpointer, whose chapter "Silent Summer" invokes a chilling reminder of Rachel Carson's predictions made 25 years ago in *Silent Spring*, and from the editorial skills of Sybil Wong, Philip Friedman and Michael Tanzer, who helped with the early drafts of the manuscript. Thanks also go to John Oakes and Dan Simon of Four Walls Eight Windows for publishing the book.

We are grateful for the generous financial support of the CS Fund, Deer Creek Foundation, Environmental Research Foundation, W. H. and Carol Ferry, Friedson Philanthropic Funds, Fund for Investigative Journalism, John I. Kennedy, Gloria White McNally, National Community Funds, Public Concern Foundation, Alexander C. Stewart, and the Youth Project. We also wish to acknowledge the help of the Council on Economic Priorities and the Institute for Policy Studies, which have sponsored portions of our environmental research over the past four years.

Special thanks go to the Reverend Benjamin Chavis and Charles Lee of the United Church of Christ Commission for Racial Justice, which has sponsored the Radiation and Public Health Project and its efforts to analyze the way in which low-level radiation disproportionately affects low-income minority groups. In late 1988, we joined efforts with CRJ in creating the RPHP in order to extend our studies of the health effects of nuclear releases, and to stimulate public debate on these controversial issues.

JAY M. GOULD
Director

BENJAMIN A. GOLDMAN
Associate Director

Radiation and Public Health Project
New York, December 1989

LIST OF TABLES AND FIGURES

CHAPTER 1

OVERVIEW

Radiation released by nuclear technologies has had a fearsome effect on the environment and human health.

Since the atomic bomb attacks on Japan in 1945, considerable research has focused on the health effects of radiation. Early studies examined the survivors of Hiroshima and Nagasaki. Subsequent laboratory experiments studied analogous kinds of whole-body irradiation. The conventional wisdom from this substantial body of research is that high doses of radiation caused by bomb blasts can seriously injure human health, but that small doses of radioactive fallout, often called "low-level" radiation, do little harm.

There is now reason to fear that low-level radiation from fallout and from nuclear reactors may have done far more damage to humans and other living things than previously thought, and that continued operation of civilian and military nuclear reactors may do irreversible harm to future generations as well.

The chief findings in this book revolve around statistical estimates of excess deaths that have never before been part of the public debate on the dangers of low-level radiation. They may shock the general reader, because there has been a sustained effort to withhold official data from the public, as discussed in Chapter Six.

Nuclear physicists realized as early as 1943 that fission products released into the atmosphere could enter into the food chain, and, when ingested, could accelerate the deaths of millions of persons worldwide. As related in Chapter Seven, Linus Pauling and Andrei Sakharov calculated in 1958 that millions of people would die prematurely from the ingestion of fission products resulting from fallout from atmospheric bomb tests.

Today we are in a position to review the official U.S. mortality statistics for nearly nine decades of the twentieth century, to find that the chilling predictions of Pauling and Sakharov may have been fulfilled not only during the period of atmospheric bomb tests but after every major accidental release of nuclear fission products.

Most previous studies of the health effects of low-level radiation were based on theoretical extrapolations of how many cancer deaths can be expected from exposure to high-level radiation, taken from the experience of the Hiroshima and Nagasaki victims. In this book, we take a completely different, pragmatic approach, guided by the pioneering work of radiation physicist Ernest Sternglass and physician and epidemiologist Carl Johnson. We analyze the mortality data collected from official death certificates filed in the wake of large radiation fallouts. In this way we can estimate the dose response to low-level radiation after the fact, rather than as a matter of theoretical speculation.

As statisticians, we define an "excess" number of deaths at any time and place as the difference between the number of deaths actually observed and the number that would be expected based on national norms, when that difference is too great to be attributed to chance (as detailed in the methodological appendix). We have found that releases of low-level radiation from nuclear power and weapons plant reactors have consistently been followed by large numbers of "excess" deaths.

Chapter Two deals with perhaps our most startling discovery, that radiation from the April 26, 1986 Chernobyl disaster, which reached the U.S. early in May of 1986, was followed almost immediately by an extraordinary force of mortality, amounting to perhaps 40,000 excess deaths in the summer months, especially in the month of May. The acceleration in deaths particularly affected the very young, the very old, and those suffering from infectious diseases such as AIDS, suggesting that the ingestion of Chernobyl fission products had an immediate adverse impact on those with vulnerable immune systems.

The Chernobyl disaster released so large a volume of fission products into the atmosphere so quickly that its immediate effects, though thousands of miles from the source, were revealed by the analysis of the official monthly mortality reports of two nations that make such data publicly available—the U.S. and West Germany.

Our results were unexpected, but when we went back to examine the mortality data associated with previous large nuclear releases, we found the same pattern of excess deaths among the very young and very old. We found immediate increases in infant mortality and in total deaths (primarily comprising older persons), which were followed later by annual increases in excess cancer deaths. These excess deaths may be linked to damaged immune systems from the ingestion of fission products: in particular, radioactive iodine, which damages fetal thyroids, and radioactive strontium, which concentrates in the bone marrow.

This book can be viewed as an epidemiological "whodunnit," with the suspect revealed by Chernobyl in 1986, and the web of circumstantial evidence traced back to 1945.

Deadly Deceit has two major parts. The first part (Chapters Two through Six) presents our findings from large databases of official measures of radiation and of mortality. (The development of these databases is described in the afterword.) In the first part, statistical tests identify significant increases in mortality, controlling for alternative explanations with geographic and temporal comparison groups, and, on occasion, with tests of multiple variables.

The second part of the book (Chapters Seven through Ten) considers the implications of the case study findings, pointing the way for future studies by offering some speculative hypotheses that grow out of the case study findings.

Much of the most telling information this book offers is in its charts, which allow the reader to visualize the findings without going through the huge masses of tabular data on which they are based. We have made the detailed data available to those who are interested. One notable use of our related databases was a report by Greenpeace USA that demonstrated the toxicity of the Mississippi River. Greenpeace found that from 1968 to 1983 there were some 66,000 excess deaths in the counties bordering the river, a figure greater than the number of Americans who died in the Vietnam War. In this book, we found similarly disturbing clusters of excess deaths associated with radiation releases. A sampling follows.

Between 50,000 and 100,000 excess deaths occurred after releases from accidents at the Savannah River nuclear weapons facility in 1970 and again at Three Mile Island in 1979. The 1970 Savannah River reactor rod meltdowns were revealed in Congressional hearings held by Senator Glenn (D-Ohio), after 18 years of official concealment. Our discussion in Chapter Four of significant mortality increases in South Carolina and neighboring southern states following the 1970 Savannah River accidents is the first such analysis published. Among the many causes for the increased mortality, we found extraordinary increases in infant deaths from birth defects. The government claims that no radiation was released as a result of the accidents, yet because the Savannah River facility is under military control, accurate emissions data are not publicly available. The significant increases in excess deaths suggest a substantial release may in fact have occurred.

The Brookhaven National Laboratory has documented hundreds, if not thousands, of "routine" and accidental civilian reactor releases since the mid-Sixties, the largest of which occurred at Three Mile Island in 1979. Chapter Five describes our findings of significant mortality increases following the accident, particularly in the ten-county area closest to the stricken reactor. As in the Savannah River case, excess infant deaths from birth defects increased significantly after the Three Mile Island accident, as did excess deaths from child cancer, lung cancer, heart diseases, and other causes.

Chapter Six provides evidence of official concealment and falsification of key data on radiation and its health effects, indicating why these findings have never been made public before. On rare occasion, there has been official acknowledgement of the deep-seated political motivations for understating the potential health effects of low-level radiation. For example, William H. Taft, a U.S. State Department attorney, said in 1981:

> *The mistaken impression [that low-level radiation is hazardous] has the potential to be seriously damaging to every aspect of the Department of Defense's nuclear weapons and nuclear propulsion programs. . . . It could adversely affect our relations with our European allies.*[1]

This book offers reason to suspect that the lethal nature of low-level radiation is no "mistaken impression." The scale of potential damage was foreseen by Rachel Carson, Linus Pauling, and Andrei Sakharov, and was later supported by warnings from John Gofman, Arthur Tamplin, Alice Stewart, Thomas Mancuso, Karl Morgan, Carl Johnson and Ernest Sternglass.

We believe that the cumulative magnitude of atmospheric nuclear weapons testing may explain what has hitherto been a great epidemiological mystery, as described in Chapter Seven. In the 1950–1965 period, mortality statistics inexplicably stopped getting better, after decades of improvement going back to the discovery of antisepsis early in the 19th century. During this period, the volume of fission products released into the atmosphere was equivalent to the explosion of some 40,000 Hiroshima bombs, according to a thorough examination of seismic records conducted by the Natural Resources Defense Council. This terrifying figure was known to the leaders of the Soviet Union, who were responsible for two-thirds of the total yield (most of which occurred in 1961 and 1962), and to Presidents Eisenhower and Kennedy. Although the magnitude of this nuclear orgy was not publicized at the time, it led to the U.S.-Soviet agreement to ban atmospheric bomb tests in 1963, after which mortality rates resumed their annual, though somewhat diminished, improvement.

Chapter Seven also offers disturbing evidence that many

members of the baby-boom generation, who were born into the nuclear age, sustained an observable degree of immune-system damage. The successive cohorts of persons born since 1945, who were exposed to ingested fission products in utero, at birth, or in early childhood, are now registering ominous increases in their mortality rates. These generations are disproportionately affected by a wide range of immune deficiency diseases, including AIDS, Chronic Epstein Barr Virus (known as "yuppie influenza" or "Chronic Fatigue Syndrome") and many others.

Chapter Ten explores a heretical hypothesis, first advanced by radiation physicists Ernest Sternglass and Jens Scheer, that may explain why AIDS first emerged in the wetlands of Africa in 1980. These areas of high rainfall received the heaviest fallout during the years of atmospheric bomb tests, as indicated by United Nations surveys taken at the end of the 1950s. The surveys showed human bones there had the world's highest concentrations of radioactive strontium-90. The hypothesis links the damaged immune systems of persons reaching their peak years of sexual activity in the 1980s to viruses mutated by radiation, and to dietary factors such as calcium intake. Sternglass and Scheer cite the extraordinary case of the West Indies island of Trinidad, which has a population of largely African and Asian origin, and where AIDS is found only among people of African origin and not at all among those of Asian descent. This discrepancy, Sternglass and Scheer propose, may be due to the dominance of calcium-rich fish and rice in the Asian diet, which offsets the tendency for radioactive strontium (which has a chemical structure similar to calcium) to concentrate in their bones.

Chapter Nine examines the potential consequences of huge emissions from the Millstone reactor in Connecticut in 1975, the second largest civilian accident after Three Mile Island. These emissions may have set off a cancer epidemic centered in the two neighboring counties of Middlesex and New London that still continues. In our effort to investigate this epidemic at the local level, we found that publication of cancer mortality data by township, routinely available from the Connecticut Department of

Health Services since the 1930s, was terminated in 1977. We think that post-1976 mortality and morbidity data for the towns close to the Millstone reactors may also throw light on the outbreak of Lyme disease, first reported in the area near Millstone during the fall of 1975.

An equally startling hypothesis is posed in Chapter Eight, where we suggest that fresh milk from dairy farms near nuclear reactors may have contributed, along with increasing poverty and other causes, to deteriorating infant mortality in certain urban areas over the past two decades.

This hypothesis was suggested to us by statistics related to the Nuclear Regulatory Commission's unprecedented closing of the Peach Bottom reactors on the border of Pennsylvania and Maryland on March 31, 1987 because of operator negligence going back to 1974. The reactors had a long record of excessive releases of the short-lived radioactive element iodine-131. The Peach Bottom reactors are located in one of the nation's most productive dairy farming areas, which supplies fresh milk to the entire Mid-Atlantic area, including the cities of Baltimore and Washington, D.C. After Peach Bottom was closed, in the summer of 1987 infant mortality in Washington, D.C. plunged to the best rates in some 20 years.

We then found a statistically significant correlation between changes in infant mortality over the past two decades and regional risks of exposure to milk contaminated from civilian reactor emissions since 1974. The fourteen states in the Midwest and Mid-Atlantic regions with the greatest risk of exposure to contaminated milk also had the worst infant mortality performance. Analyzing these data, we found that while the exposure risk of eight Midwest states was 440 times greater than that of three northern New England states, the corresponding infant mortality performance was only ten percent worse. This evidence suggests that the dose-response is "supralinear" rather than linear, which means that infant mortality rises more rapidly at low doses.

Another example of the supralinear relationship was offered in the wake of Chernobyl. The June 1986 increase in infant deaths over June 1985 in the U.S. was a full ten percent of the increase in West Germany's Baden-Württemberg province, even though U.S. radiation levels were only one-hundredth to one-thousandth as great.

This crucial evidence supports the 1972 laboratory findings of Dr. Abram Petkau, a Canadian radiation biologist, on the dangerous effects of "free radicals" created by exposure to low-level radiation. Free radicals are charged particles that can penetrate and destroy the blood cells of the immune system, especially at low-levels of radiation.

Our findings of a supralinear effect also agree with similar findings for cancer mortality from exposures to low-level radiation made by four eminent authorities: Dr. John Gofman, Dr. Karl Z. Morgan, Dr. Thomas Mancuso and Dr. Alice Stewart. All four scientists worked at various times for the U.S. Atomic Energy Commission or Department of Energy. All four concluded that the dose-response relationship was supralinear, which means that there is no level of radiation low enough to be deemed "safe". The government terminated the services of all four when they each, independently, came up with what Dr. Gofman has called the "wrong" answer—that is, the opposite of what the AEC wanted to hear.

The supralinear dose-response for infant mortality may apply to all deaths from immune-system damage caused by radiation-induced free radicals (the so-called "Petkau effect" which is discussed in the methodological appendix). This generalization is supported by a projection of the current trend in the U.S. age-adjusted mortality rates (see Chapter Seven). This projection suggests that without fundamental change, the death rates of all age groups will begin to rise in the 21st century, cancelling out previous advances in longevity.

One striking number should be mentioned here: The statistical probability is less than one in one million that during the summer following the Chernobyl accident the excess deaths observed in the U.S. were due to chance. Equally improbable were the excess deaths observed in West Germany during the same time period. And, as related in Chapter Three, ornithologist David DeSante found at the same time that the number of newly-hatched land birds counted by the Point Reyes Bird Observatory in California in the late spring and summer of 1986 dropped 62 percent below the average of the preceding decade. The probability that the simultaneous summertime mortality peaks in the U.S., West Germany, and among birds are unrelated random events can be expressed mathematically as one out of 10^{30}, that is, one out of:

1,000,000,000,000,000,000,000,000,000,000.

Even so, we acknowledge that something other than radiation or chance may have caused the unusual mortality phenomena in the summer of 1986, as well as the other significant mortality increases associated with radiation releases described in this book. Our evidence is largely statistical and as such not one-hundred percent conclusive, but these significant statistics cannot be ignored. This book is a challenge to the scientific community to identify plausible alternative explanations.

The charges made here are too important to be left to the experts for resolution. Continued reliance on nuclear technologies may pose an on-going threat to life on earth. The potential danger warrants the widest possible audience and public debate. As the final chapter indicates, it is not too late to eliminate the chief sources of radioactive pollution. We can cite as hopeful examples Wyoming and Montana, two states far from nuclear emissions, where infant mortality rates are among the best anywhere in the world today.

CASE STUDIES

CHAPTER 2

FALLOUT FROM CHERNOBYL

Prior to the nuclear reactor accident in the Ukraine on Saturday, April 26, 1986, very few people in the world knew the name "Chernobyl." But on April 29, when a United States surveillance satellite spotted Chernobyl reactor Number Four burning red with fire, confirming rumors from Sweden that a dangerous nuclear accident had occurred, "Chernobyl" became infamous overnight, especially to the more than 20 nations that were in the fallout's path.

Now regarded as the world's worst nuclear accident, the Chernobyl disaster was caused by "unbelievable" errors in judgment by the plant's technicians, according to the chairman of the Soviet State Committee on Atomic Energy Use.[2] Ironically, shortly before the accident, Soviet nuclear scientists had stated that a catastrophic accident at the Chernobyl plant was "impossible."

But the impossible happened. At 1:23 a.m. that fateful Saturday morning, a thunderous blast lifted the massive concrete lid from the reactor and released a plume of radioactive debris that was carried two thousand meters into the air. The initial explosion split the reactor core and set fire to the surrounding buildings. The reactor core burned for two more weeks, releasing radioactive contaminants all the while. Within a few days, according to the Lawrence Livermore National Laboratory, hundreds of millions of curies of radiation were released into the biosphere. This release may have amounted to about one-tenth of the nuclear fission products that had been spread by all bomb tests since 1945.

This chapter explains how the Chernobyl accident may signal the true danger of low-level radiation to all forms of life. The evidence offered here, that the Chernobyl fallout was much more lethal than has been officially acknowledged, is drawn from many disciplines in addition to statistics, including biochemistry, medicine, radiation physics, epidemiology, and ornithology.

Evidence of Chernobyl's unexpected health impacts first came at a May 1987 conference on the effects of radiation on health, held in Amsterdam by the World Information Service on Energy. Chilling stories from many parts of Europe were told there of the effects of high levels of radiation from Chernobyl, including accounts of human and animal miscarriages. Present at the conference were two Soviet physicians who described the heroic efforts to save the lives of Chernobyl firemen who had been exposed to high-intensity radiation. But they professed to have no information about the health effects of low-level radiation. Two physicians from Cracow, on the other hand, horrified the audience by announcing their belief that live births in Poland experienced a staggering drop after Chernobyl. However, they did not make clear whether these were due to spontaneous or induced abortions after the arrival of the Chernobyl fallout.

Subsequent inquiries to the Polish Embassy in Washington elicited the information that the number of live births in Poland during 1986 dropped ten percent below the 1985 level. If there was little change in the first four months of 1986 prior to the Chernobyl accident, then there would have been a drop of 14 percent after the Chernobyl radiation arrived, and perhaps as much as a 20 percent drop during the summer months. These indications of dramatic infant mortality effects in Europe raised the question of whether sufficient radiation had reached the U.S. to produce detectable adverse health effects.

On May 5th, nine days after the Chernobyl accident, monitoring stations in the State of Washington—9,000 miles from the Ukraine—found radioactive iodine-131 in the rainfall, with test stations around the state reporting peak values between

May 12th and 19th. The earliest readings in Richland and Olympia recorded iodine-131 concentrations in rainwater of approximately 170 picocuries per liter (pCi/l). The highest levels in the Pacific Northwest were found in Spokane, where they peaked at over 6,600 pCi/l by May 12th. By May 16th, low-level radiation was recorded by about 50 Environmental Protection Agency milk monitoring stations in states that received the "Chernobyl rain." No warnings against drinking the milk were issued by public health authorities because the reported levels were regarded as safe. Extensive fallout data were subsequently published by the EPA.[3]

Government data indicate that there were also statistically significant increases in deaths in the U.S. during May of 1986.[4] These statistics showed a surprising 5.3 percent increase in the total number of deaths in the U.S. in May 1986 over the same month in the previous year. This was not only statistically significant (with a probability of less than one in a thousand of being a chance event); it was, in fact, the highest annual increase in May deaths recorded in the U.S. in 50 years. There were also high percentage increases in deaths in the three succeeding months.

Figure 2-1 shows the daily pattern of radioactive iodine found in fresh milk at a milk station in the New York-New Jersey metropolitan area during May 1986, as recorded by the Department of Energy's Environmental Measurement Laboratory. Because the radioactivity of iodine-131 decreases quickly (it has a half-life of only eight days), the peak occurred in mid-May. Other radioactive isotopes such as cesium-137, strontium-89, strontium-90, and barium-140 were also identified.

Figure 2-2 shows that states comprising the South Atlantic region registered a massive 28 percent increase in infant mortality in June of 1986 over June of 1985. In the U.S. as a whole, infant mortality jumped 12.3 percent over the previous June.[5] The infant mortality rate, defined as the number of babies dying in the first year per 1,000 live births, is one of the most sensitive indicators of monthly changes in public health.

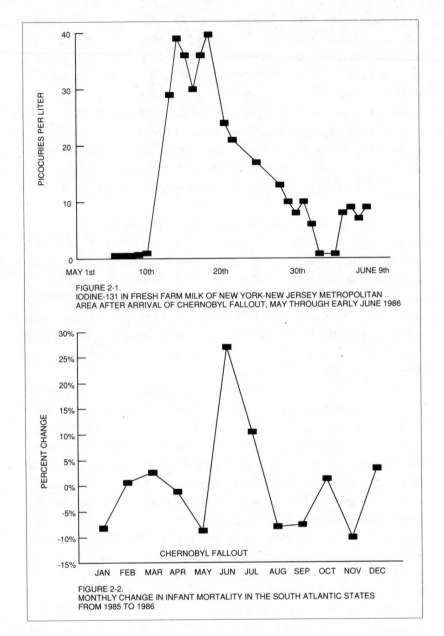

FIGURE 2-1.
IODINE-131 IN FRESH FARM MILK OF NEW YORK-NEW JERSEY METROPOLITAN
AREA AFTER ARRIVAL OF CHERNOBYL FALLOUT, MAY THROUGH EARLY JUNE 1986

FIGURE 2-2.
MONTHLY CHANGE IN INFANT MORTALITY IN THE SOUTH ATLANTIC STATES
FROM 1985 TO 1986

Figure 2-3 records statistically significant increases in the number of deaths in May 1986 compared to May 1985 for two age groups (young adults aged 25 to 34 years old, and elderly persons over 65), and for three causes of death (infectious diseases, AIDS-related diseases, and pneumonia). The probability that any of these changes is the result of chance is less than a one in a thousand, so the likelihood that all could simultaneously be a chance result is less than one in one million.

The latest National Center for Health Statistics estimate of the number of deaths in the four summer months of 1986 is 674,000— a 2.5 percent gain over 1985. Because of the large numbers of people involved, this is also a statistically significant increase, with a chance probability of less than one in one thousand. By September of 1986, most of the immediate mortality effects appeared to have diminished.

The data offer strong evidence for a relationship between the measured radioactivity in milk and the changes in monthly mortality for adults and infants across the country. An examination of the changes in total deaths among the nation's nine census regions for May-August 1986, compared with 1985 numbers, showed a high correlation with levels of radiation in pasteurized milk as reported by the EPA.

Figure 2-4 indicates that the higher the level of radioactive iodine found in milk in a region, the higher the percent increase in total deaths. The points in Figure 2-4 show the percent increase in total May deaths at peak iodine-131 concentrations in each region.[6] The area with the least rainfall in May 1986 and the lowest radioactive iodine concentration in milk (17 pCi/l) was the West South Central region (consisting of the states of Texas, Arkansas, Louisiana, and Oklahoma), which registered no change in mortality. On the other hand, the Pacific region, mainly California and Washington, had the highest concentration of iodine in milk (44 pCi/l) and registered the highest increase in total deaths.

The observed data trend is best represented by the curving solid line, which is a "logarithmic" fit, meaning that at higher levels of

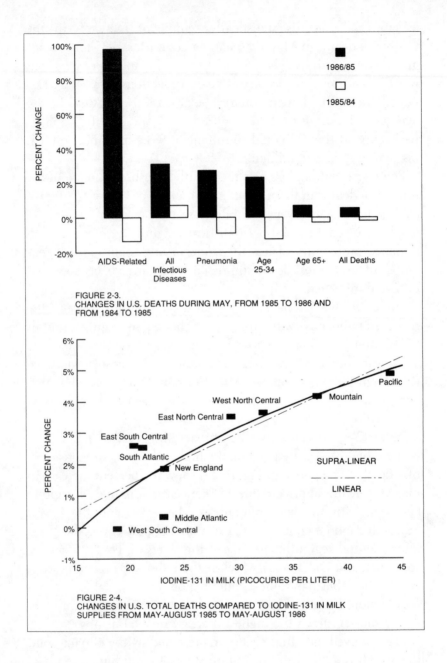

FIGURE 2-3.
CHANGES IN U.S. DEATHS DURING MAY, FROM 1985 TO 1986 AND
FROM 1984 TO 1985

FIGURE 2-4.
CHANGES IN U.S. TOTAL DEATHS COMPARED TO IODINE-131 IN MILK
SUPPLIES FROM MAY-AUGUST 1985 TO MAY-AUGUST 1986

radiation exposure, the rate of the mortality response diminishes. The dotted line in Figure 2-4 is a straight "linear" fit to the data, which predicts that deaths increase constantly as the radiation dose increases.

Figure 2-5 illustrates the logarithmic nature of the dose-response function, taken from the previous figure, and extends it to the higher concentrations of radiation measured in Europe in the wake of the Chernobyl accident. It indicates that deaths increase rapidly with iodine-131 levels below 100 picocuries per liter, but the percentage increase flattens out at higher radiation levels. The dotted line represents the relationship between dose and adverse mortality response that physicists and health officials have assumed to be true since Hiroshima. As can be seen from the figure, the conventional dose-response assumption that has been derived by extrapo-

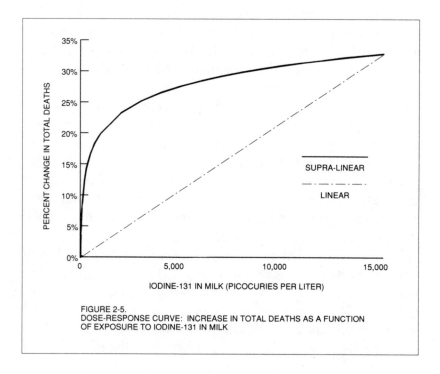

FIGURE 2-5.
DOSE-RESPONSE CURVE: INCREASE IN TOTAL DEATHS AS A FUNCTION OF EXPOSURE TO IODINE-131 IN MILK

lating from high doses greatly underestimates the effect at low radiation intensities.

This assumption appears to have been wrong because it was based on the experience of short bursts of high radiation intensities, as in the case of medical therapeutic uses, or flashes from an atomic bomb. These effects act primarily on the DNA in the cell, and such damage can be repaired effectively by enzymes. This process is totally different from the indirect, immune-system-impairing "free radical" oxygen (O_2^{-1}) mechanism which dominates radiation damage at very low dose rates.[7] (For more details, see the methodological appendix discussion of the "Petkau effect" as a causal factor that may account for the major mortality effects of low-level radiation.)

It was the deaths of infants, young adults suffering from infectious diseases, and elderly persons that increased greatly in the summer of 1986. These population groups share the common characteristic of having relatively vulnerable immune systems. Any additional insult to such immune systems can hinder the response to ailments and stress. Affected elderly people were probably already ailing; the Chernobyl radiation may have further weakened their immune systems' resistance.

Perhaps the most sobering statistic concerns young adults who suffered unusually high increases in deaths. Immune system damage appears to be behind the significant increase of deaths from pneumonia and infectious diseases, especially those related to AIDS, which peaked in May 1986 and the following summer months. Normally, summer months show the lowest death rates from infectious diseases, because epidemics from influenza and the like usually occur during the winter.

If the Chernobyl fallout is responsible for these steep and highly unlikely mortality increases, then this is the first evidence using large populations that suggests the dose-response curve at very low dose rates of fallout radiation exposures is logarithmic and not linear, contrary to generally-accepted assumptions. The medical and scientific community has long believed, on the basis of linear extrapolations from high doses, that low-level radiation from fall-out and nuclear plant releases can be dismissed as posing a negligibly small danger. The Chernobyl experience indicates that this assumption may underestimate the effect of low radiation doses for the most sensitive members of the population by a factor of about one thousand.

These Chernobyl findings were reviewed by Drs. Donald Louria and Marvin Lavenhar, of the Department of Preventive Medicine and Community Health of the New Jersey Medical School. Despite their initial skepticism, after two months of review they could find no errors in the calculations or plausible alternative explanations. The findings were made public at the First Global Radiation Victims Conference in New York on September 1987, and were ultimately published in an article by Jay Gould and Ernest Sternglass by the American Chemical Society in the January 1989 issue of *Chemtech*.[8]

After the initial presentation of the U.S. mortality findings in *Chemtech*, a critic observed that if these results could be attributed to the Chernobyl radiation, "people in Europe should have died like flies" since radiation levels there had been so much higher. The answer to this soon came, with further confirmation of the logarithmic nature of the dose-response curve, from Professor Jens Scheer, of the University of Bremen (West Germany), whose collaborators Michael Schmidt and Heiko Ziggel had assembled monthly mortality data from areas in West Germany that were heavily exposed to radiation from Chernobyl.[9]

Figure 2-6 shows that increases in infant mortality peaked in June 1986 in Baden-Württemberg, as had also occurred in the U.S. But in contrast to the 12.3 percent increase in the U.S. over June of 1985, the corresponding gain for the more-heavily irradiated West German province was 68 percent. That was the highest increase registered in all of West Germany. In the less-heavily irradiated North the effect was much lower. This is a strong indication that the increase in the South was due to radiation and not to some other hypothetical factor.

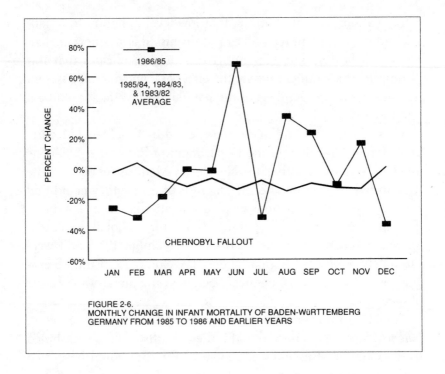

FIGURE 2-6.
MONTHLY CHANGE IN INFANT MORTALITY OF BADEN-WüRTTEMBERG
GERMANY FROM 1985 TO 1986 AND EARLIER YEARS

According to Dr. Scheer, "particularly conspicuous were babies that died in the first week of birth, and if you look more in detail you will see there were also conspicuous rises in the infant mortality for those whose conception occurred right in the first week of Chernobyl as well as those in their last month of gestation, which were born in the summer of 1986."[10] He also noted that West Germany experienced small but statistically significant increases in Down's Syndrome among children conceived during May 1986. In addition, he reported a marked increase in the use of a German health insurance company compared to the previous ten years with respect to allergic diseases.

The National Center for Health Statistics has reported that 1986 was also the peak year in the U.S. for both incidences of acute medical conditions and the number of disability days associated with such illnesses during the period 1982–87.[11]

Radiation levels in Europe were one hundred to one thousand times greater than in the U.S., but the summertime increase in European infant mortality was only about ten times higher than in the U.S. This is further evidence for the logarithmic nature of the dose response curve for low-level radiation (although warnings not to drink milk were widely heeded in Europe, which probably reduced exposure to the fetuses).

At first, the American media completely ignored the U.S. mortality findings, although they were reported on the front pages of leading Japanese and Canadian newspapers in September and October of 1987. By January 1988 the Chernobyl data became the subject of an editorial commentary in the London *Independent*, entitled "Longer arms for the grim reaper: new American research has alarming implications." This was soon followed by "One deadly summer" in the London *Economist* (January 30, 1988), which was carried as a syndicated story in dozens of major U.S. newspapers, and by a detailed story in the Italian news weekly *Il Mondo*, "Do you remember Chernobyl?" After the February 8th publication of an article on the subject in *The Wall Street Journal*, many more stories were finally published in the U.S., as well as in other countries.

The publicity prompted some interesting comments by leading health authorities. *The Wall Street Journal*, for example, quoted Dr. Louria of the New Jersey Medical School: "You cannot look at this blip in the data and say 'so what'. . . . I've been persuaded that there is enough there to merit a good look. It would be unwise to treat Gould's findings dismissively and equally unwise to overinterpret them."[12]

Some guarded comments came from federal health officials. Neal Nelson, an EPA radiation biologist, was quoted by *The Wall Street Journal* as saying, "the incremental radiation doses from Chernobyl were only a tiny fraction of ordinary background radiation levels. To say that such an increase could trigger such a change in immune susceptibility does not seem reasonable . . . but the work [of Gould and Sternglass] should be evaluated on its merits."[13]

Sharon Ramirez, a spokeswoman for the National Center for Health Statistics (NCHS), told the *Seattle Times* that the mortality figures were correctly taken from federal publications and that, "we don't make a 2,300-death mistake; we're usually right on the money." In later editions, this quotation was replaced by a statement from Patricia Starzyk, a Washington State Department of Social and Health Services investigator, who said, "I don't know where they got their numbers from, but their numbers are wrong. . . . I can only guess there was some error in the publication they were using." Rather than clarifying the discrepancy between the two statements, the *Seattle Times* quoted an August 1986 report of the Washington State Office of Radiation Protection, which said, "the immediate health and safety of the citizens of the state were at no time jeopardized by fallout from the Chernobyl accident."[14]

Dr. Harry Rosenberg, chief of mortality statistics at NCHS, was quoted by the *Toronto Globe*: "one cannot *a priori* dismiss these data. You must look at them very carefully and hope that Gould and Sternglass shed some light on the problem of radioactivity."[15]

Comments from nuclear officials were far more critical. Warren Sinclair, of the government-supported National Council on Radiation Protection, was quoted by the American Medical Association's *Medical News* as saying: "I don't place much credibility on anything involving the sort of low-level radiations that were present in this country after Chernobyl. They are unlikely to cause changes such as those in the statistics."[16]

After several of the articles appeared, we received a letter from Dr. David DeSante, a researcher at the Point Reyes Bird Observatory in California, which stated:

> We documented a massive and unprecedented reproductive failure of most species of landbirds at our Palomarin Field Station [located 25 miles north of San Francisco] during the summer of 1986. The number of young [newly hatched] birds captured in our standardized mist-netting program was only 37.7 percent of the pre-

vious ten-year mean ... Interestingly, the reproductive
failure did not begin at the start of the breeding season
but only after about one month of the season had passed,
that is, for birds hatched about mid-May ... Further-
more, there seemed to be a slight recovery of reproductiv-
ity very late in the season ... Might this implicate
iodine-131?[17]

Figure 2-7 summarizes Dr. DeSante's findings. It shows a drop in
the number of newly hatched landbirds counted from May 10 to
August 17, 1986. Once again, the probability of the reproductive
failure he found being the product of chance is infinitesimal.[18]
DeSante's findings indicated that the observed impacts on human
mortality were replicated in reproductive and immune system
failures among birds during the exact same time period—mid-May

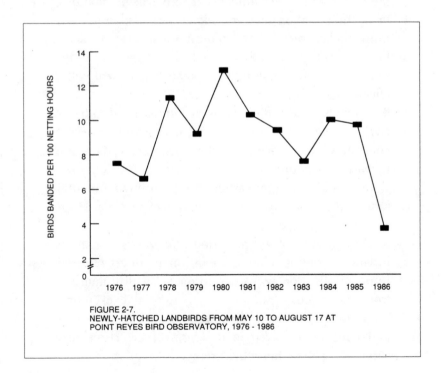

FIGURE 2-7.
NEWLY-HATCHED LANDBIRDS FROM MAY 10 TO AUGUST 17 AT
POINT REYES BIRD OBSERVATORY, 1976 - 1986

to mid-August of 1986. (The next chapter provides a fuller description of DeSante's findings.)

The Chernobyl accident, for all its tragic consequences, offers an
opportunity to consider whether the price for continued operation
of nuclear reactors is too great for society to pay. The following
special characteristics of the Chernobyl accident makes such a
reassessment imperative:

1. It involved a much *larger exposed population* than
any earlier study.
2. It involved a *normal population*, not hospital patients, workers of a limited age range of 18–65, or survivors of a traumatic event such as the destruction of
Hiroshima and Nagasaki.
3. It involved *extremely low doses* of radiation, comparable to those received from distant nuclear detonations,
or to the radiation now considered permissible releases
from nuclear reactors and plutonium separation facilities. There is, therefore, no need to make any assumptions as to how best to extrapolate theoretically the
effects expected at very low doses.
4. It involved *accurately measured amounts of radioactivity in the diet*, including milk, over a wide range of
concentrations, especially when European data are
taken into account. No such accurate measures of dose
were available in any earlier studies of environmental
radiation exposures or direct radiation exposure at Hiroshima or Nagasaki.
5. It involved *internal radiation exposure* as a result of
inhalation or ingestion from fission products released
from reactors or bomb explosions—as distinguished
from the external radiation from normal background
levels, radon exposures, diagnostic or therapeutic X-rays
and neutrons involved in the cases of the Hiroshima
survivors or the Chernobyl firemen.

Thus, the statistics emerging from the Chernobyl disaster permit, for the first time ever, the establishment of a dose-response relationship at extremely low doses in a normal human population, down to a small fraction of the doses encountered in the normal environment. This relationship could be tested for its predictive value by comparing the mortality effects in the United States with those in Europe, where the population was exposed to much greater doses.

The data show that protracted internal exposures at low doses do not lead to a reduced effect, but rather to an increased marginal effect, as compared to brief but high exposures. They show also that there is no safe threshold for small exposures comparable to those from normal background radiation sources, and certainly there is no "beneficial" effect as has recently been argued by nuclear proponents. Normal background radiation that enters the food chain always has been a human health risk. But the dangers of low-level radioactivity from manufactured fission products are relatively new.

Technically speaking, calculations performed on the post-Chernobyl health data show that the dose-response curve at low doses is neither a quadratic, upward curving one, nor a straight line (linear) relationship. It is rather a supra-linear or logarithmic function that rises more rapidly at low doses than at high doses. This logarithmic form of the dose-response curve is consistent with the laboratory results of Petkau and others on the indirect, free-radical mediated effect of radiation on cell membranes, particularly the oxygen "free radical" now known to be involved in a wide range of immune deficiency diseases (see the discussion of this research in the methodological appendix).[19]

The logarithmic form of the dose-response also means that there is actually no conflict between the new evidence about the serious effects of very low, protracted exposures and the results of earlier studies on high dose-rate exposures in laboratory animal and atomic bomb survivors. The effect per unit dose is much lower at high intensity and short exposures than at extended low dose-rate exposures. This may be due to the lower efficiency of free radical damage at high intensities, when large numbers of free radicals tend to negate each other.

Because similar effects were discovered independently for birds, there is reason to believe that the free-radical effect of low-level radiation may affect other animals and perhaps even plants.[20] The special vulnerability of young adults, however, is particularly disturbing. Chapter Seven posits that young adults suffered so large a mortality toll in 1986 because they were born during the peak years of atmospheric bomb testing in the 1950s and may have sustained damage to developing immune systems sufficient to make the immediate effects of Chernobyl puny in comparison.

CHAPTER 3

SILENT SUMMER*

Twenty-eight years ago, ecology pioneer Rachel Carson warned in her prescient book *Silent Spring* that unless humanity were to stop polluting the biosphere with chemical and radioactive poisons, some future spring would yield "only silence. . . over the fields and woods and marsh."

Carson's imaginary spring drew uncomfortably nearer reality in 1986, when a massive and unprecedented decline in landbird productivity was witnessed and documented by ornithologist Dr. David F. DeSante at the Point Reyes Bird Observatory (PRBO), located about 25 miles north of San Francisco.

"Usually," DeSante said, "there's a lot going on when you walk down the net lanes in July. There are flocks of 'punks' [juvenile birds] and family groups of bushtits. Juvenile sparrows are collecting in little groups, and warblers are flying through the trees. The young birds are squeaking and chirping and some of the adults are singing." [21]

But when he walked the net lanes on July 22, 1986, there was a striking change. Instead of the exhilarating songs of multitudes of adult birds involved with their breeding and nurturing activities, and the squeaks and chirps of the fledgling young, he met an ominous silence.

---

* THIS CHAPTER WAS WRITTEN BY KATE MILLPOINTER.

"There just were no young birds," he said, "and the adults had stopped singing. I guess they had just given up."

For more than a decade DeSante, who earned his doctorate in biological sciences at Stanford and has a master's degree in engineering, headed a standardized mist-netting and banding project at PRBO, the landbird biomonitoring program. The young birds get caught in the fine filaments of the mist nets, and then are counted and banded for tracking purposes. Unlike most researchers, who usually concentrate on just one species, DeSante monitored the reproductive success of 51 species of landbirds, which makes him a big-picture ornithologist.[22]

Generally, most banding programs are conducted during the fall and winter, when birds are migrating. However, DeSante conducted his research during the spring and summer breeding season, when thousands of birds nest at PRBO's Palomarin Field Station, located just inside the southern end of the Point Reyes National Seashore in Marin County. As such, his work was unique in North America.

The breeding season had started out auspiciously enough in 1986, and by May it seemed clear that the season was going to be better than usual. Based on the rather high amounts of rainfall California had received that winter, DeSante and his researchers expected landbird productivity to be ten percent above normal. In fact, during the first 30 days of the monitoring period—May 10 through June 8—the capture rate was almost twelve percent above normal.

Then in mid-June, during the fourth of ten ten-day monitoring periods, the researchers observed that the number of birds netted was only 56 percent of the previous ten-year average. Although this was a lower reproductive rate than normal, it is not unusual to see a decrease during that part of the breeding period, according to DeSante. So the researchers dismissed this early indication that something was amiss, expecting to see a rapid improvement.

The improvement never came. Instead, the numbers got worse— almost on a daily basis. By the eighth monitoring period, which occurred in late July, productivity dropped to 24 percent of normal. And this during a time when peak numbers of birds are usually captured. From 1976 through 1985, the average daily capture for July had been more than 30 birds, and 60 and even 90 bird days were common, according to DeSante. But in July 1986, no more than 24 birds were netted in any single day, and there were days when only three birds were captured.

Dismayed by these results, DeSante and his colleagues began an arduous seven-week computer analysis of the captures of newly-banded birds for the years 1976–1986, hoping that the data they generated would provide a clue to the mysterious decreases of young birds. They ruled out pesticides, herbicides or other chemicals, since no applications were known to have occurred in the past eleven years within at least two kilometers of the area.[23] And starvation was evidently not a factor, because the food supply was plentiful relative to recent years.

"Nobody could think of anything to explain this," DeSante said. "So I said, as a joke, 'Well it must have been Chernobyl,' and everyone just burst out laughing. Because when that fallout cloud passed over, and when it rained, the radio reports said that there was no reason to worry—and no reason to even wash the vegetables and fruit—the amount of radiation is insignificant—don't get alarmed—everything is fine. So we didn't think about it anymore."

Acting on the hunch that he was not the only researcher witnessing the plummeting bird populations, DeSante called Dr. Donald L. Dahlsten, at the University of California. Dahlsten has conducted nesting-site, reproductive and life-span studies on mountain and chestnut-backed chickadees at two study sites: Blodgett Forest (since 1972), located in the western Sierra Nevada; and Modoc County (since 1964) in northeastern California, about 350 miles from Sacramento.

Formerly an executive editor of *Environment*, Dahlsten is professor of entomology and until recently chaired the Division of Biological Control at the University of California in Berkeley. He specializes in how birds control forest insects, with particular emphasis on the disruption of this natural balance caused by over-use of pesticides. Instead of mist-netting juvenile and adult birds, Dahlsten and his coworkers study and band the nestling or baby birds while they are still in the nesting boxes.

When asked by DeSante about how his chickadees were doing, Dahlsten said that Blodgett Forest had been a disaster that year and he did not know why.

"We noticed something was wrong as soon as we saw the first nests," Dahlsten said. "It was one of those black and white things. We were aware that there was a helluva mortality and we could not figure it out. It was the first time I had seen such a failure."[24]

When Dahlsten tallied the 1986 reproductive failures, and compared the results with previous years, he discovered that complete nest failures were at a 15-year high at Blodgett Forest, as were nestling and egg mortality.[25] Once again, pesticides and starvation were ruled out as possible explanatory factors in the unprecedented mortality spikes.

Dr. C. J. Ralph made similar observations at the Lamphere-Christiansen Nature Preserve, North of Eureka, California. Dr. Ralph witnessed a 60 percent decrease in newly-hatched White-crowned Sparrows, compared to the previous four years. An ornithologist and research scientist with the U.S. Forest Service, and adjunct professor at Humboldt State University, Dr. Ralph had independently studied the breeding biology of white-crowned and song sparrows since 1982.

"We don't know if there was unusual mortality, or lack of breeding success, but we didn't have as many juveniles to band in 1986," he remarked late in the summer of 1988.[26] "Our data are nothing in isolation. However, when added together with DeSante's work, it is interesting, because it points to a geographical component." Nevertheless, Ralph suggested that what had taken place could be "coincidence."

Researchers at the Harvey Monroe Hall Research Natural Area in the subalpine Sierra Nevada witnessed significant decreases among Oregon juncos. Nine previous years of data showed that numerous groups of juncos, with 30 to 150 birds per flock, moved up the west slope of the Sierra into the subalpine in the middle to late summer. In 1986, just a few straggling flocks of juvenile juncos were observed, with the largest group comprising only four individuals. And there appeared to be a nearly complete absence of juvenile warbling vireos and black-headed grosbeaks.

These corroborative findings convinced DeSante that the unprecedented reproductive failure was not limited to Palomarin, but had extended over much of northern California. DeSante's data also indicated that the reproductive failures occurred around May 10 or 15, because the first decreases in young were observed three to four weeks later. (Birds captured in the mist-nets are "dispersing young" that have been out of the nest for three or four weeks.) The reproductive decreases of nearly every species of landbird at Palomarin had not started at the beginning of the breeding season, but after about thirty days into it. Beginning on June 9, capture rates of young birds plunged—from 56 percent of normal, to 42 percent, to 39 percent, and finally in late July to only 24 percent of normal. Something very unusual had happened in the early part of May—but what?

Curiously enough, Dahlsten's Modoc County site, located in the far northeastern corner of California, showed reproductive numbers on the *high* side of normal. Researchers in the southern section of the state, reported the same. The explanation seemed to be associated with the heavy rain that had fallen on most of northern California on May 6, but had missed northeastern and southern California.

At this point in their investigation, according to DeSante, one of his colleagues remarked, "that is when the Chernobyl cloud was passing over," and urged that they reexamine this hypothesis. This time nobody laughed when Chernobyl was mentioned.

When DeSante and his colleagues categorized the bird species according to migratory behavior, habitat preference, and nest location, they found that the decreases were independent of those factors. However, when they classified the species according to foraging behavior, they discovered an astonishing fact—the only species not affected were woodpeckers and swallows.

At first they could not understand why the two groups of species were exempted from the decreases, but knowledge of avian diets provided a clue. They knew woodpeckers feed their young on grubs and beetles, which in turn feed on dying, dead and decomposing wood. Swallows feed their young on flying insects, which, in the vicinity of Palomarin, primarily emerge from flowing water in small creeks that contain decomposing materials.

So whatever had affected the majority of birds at Palomarin appeared to involve the primary production food chain, such as caterpillars and other larvae which eat new plant growth, and are in turn fed upon by many species of birds. Such foods are an important source of forage for warbling vireos and black-headed grosbeaks. During the entire one-hundred days of mist-netting and banding, the researchers did not net one young warbling vireo or grosbeak. DeSante believes no young were produced by those species in the vicinity of Palomarin in 1986.

By mid-September 1986, DeSante had completed a painstaking study of the combined data from Palomarin and from other areas on the West Coast. He found that while the rate of *adult* birds banded per one-hundred net hours in the summer of 1986 was eight percent below the previous ten-year mean, the rate of *young* birds banded was 62 percent below the mean. Conversely, better-than-average breeding success occurred for mountain chickadees east of the Sierra Nevada, for the subalpine community on the east slope of the Sierra Nevada, and for those in Southern California. Due to weather patterns, those areas received no Chernobyl fallout.

DeSante also found that the reproductive failures coincided, geographically, with the passage of the May 6, 1986 Chernobyl cloud over coastal Washington, Oregon and northern California. Neither past heavy spring rains, droughts, or other unusual weather conditions such as the 1982–83 El Niño winter of excessive rainfall produced such severe effects on landbird productivity as seen in the summer of 1986. Those past events resulted in only nineteen to thirty-two percent reductions in landbird productivity.

Woodpeckers and bark-gleaners—birds that feed on insects in dead and decaying wood, which absorbs no rainwater and thus no radiation—showed no decline at all. However, birds that feed on insects that feed on new plant growth, showed declines of 63 to 65 percent, and seed-eaters declined by about 50 percent. Circumstantial evidence was strong for DeSante's food-chain hypothesis.

DeSante's explanation as to how Chernobyl fallout could have spurred infant and juvenile bird mortality is based on the fact that radioactive contaminants become increasingly concentrated as they move up the food chain. A startling and disquieting example of this "transfer factor" in action, is that fish that feed on algae and ocean sediments have been found to concentrate radionuclides to levels far surpassing the amounts found in the water in which they live. DeSante suspected that iodine-131, the primary constituent found in North American fallout, was the culprit behind the reproductive failure.

The deleterious effects of iodine-131 are relatively well-documented in sheep, cattle, swine, and humans, but no comparable studies have been conducted on birds. Yet is seemed reasonable to DeSante that similar health problems might occur in birds, particularly small insectivorous ones. "Smaller birds ingest more food matter per body weight because they have a faster metabolism and so will take up a larger dose of radiation than larger birds," DeSante explained.[27]

"No others animals would be as sensitive to radiation as baby birds during their first ten days of development. Laboratory radiation studies have been done on chickens. They are very different because they hatch full-feathered and are able to run around right away. They have a long development time in the egg and they are larger and heavier in body weight. The studies that have been done on small birds such as bluebirds and tree swallows subjected them to much higher doses of radiation than Chernobyl produced. So people haven't worked with low-levels of radiation on small birds. That is what needs to be done now."[28]

DeSante wondered if a correlation could be found between the amount of radiation potentially received by the birds in various areas of the United States and their reproductive success. He decided to examine the amounts of radiation measured in pasteurized milk by the Environmental Protection Agency (EPA) across the United States, and changes in landbird numbers between 1986 and 1987, as recorded by the Breeding Bird Survey, which is conducted under the auspices of the U.S. Fish and Wildlife Service.

But birds don't drink milk, so why did DeSante use the EPA milk data? He explained it this way. When radioactive rain falls, it is adsorbed by the vegetation and concentrates on the new growth. Then it is grazed upon by first order, or primary consumers, such as cows—or caterpillars, and other larvae and grazing insects such as grasshoppers. Arboreal insectivores, small insect-eating birds that forage in trees, consume these caterpillars and other larvae, and also feed them to their young. So the amount of radiation picked up in milk is a good measure of the amount of radiation that was picked up by the primary consumers, such as grazing insects, and eaten by the birds.

DeSante further theorized that if there had been significant decreases in the reproductive success of small arboreal insectivores in 1986, the decreases should show up in the population levels of these birds in 1987, as recorded in the Breeding Bird Survey data. Indeed, he found a strong correlation between regional concentrations of iodine-131 in milk, and decreases—between 1986 and 1987—in numbers of small, arboreal, insectivorous birds. For no other birds was there a similar, significant correlation.

He suggested that by virtue of their larger body weight and lesser consumption of grazing insects, birds in other foraging categories were spared the effects of low-level radiation. He concluded that Chernobyl fallout may have adversely affected the reproductive success of small, arboreal, insectivorous birds all across the United

States, and that the severity of the effect was related to the amount of radiation that they received.

DeSante also wanted to know if survival from 1986 through 1987 differed from the previous six years for adult birds of different ages. DeSante and his researchers had seven years of survival data for three species of coastal scrub birds at Palomarin: wrentits, Nuttall's white-crowned sparrows, and song sparrows.[29] They found that in 1986–87, the survival rate of old birds of these three species was the *lowest* in seven years. In sharp contrast, the 1986–87 survival rate of one-year-old and middle-aged adult birds of these three species was the *highest* in seven years. Presumably, this was because of favorable weather conditions during the winter of 1986–87. At first these anomalous findings perplexed DeSante, because older adult birds generally survive at least as well as young adult birds.

"Older birds are generally a little bit dominant over younger birds, more experienced, better able to find shelter, and they usually have the best territories," DeSante explained. "If the problem were food supply, again the older birds generally do better, are more dominant and thus better able to get food."[30]

As a result of his careful studies of the landbird biomonitoring program at Palomarin, he had discovered that there was a reproductive failure that affected young or embryonic birds. By the time the fledglings were out of the nest, there were sixty-two percent fewer than there should have been. The survival data demonstrated that the very old were affected as well.

Finally, DeSante wondered if the survival of one-year-old birds in 1987 would provide an indication of the magnitude of the reproductive failure in 1986. Again, he compared the number of young birds of the same three species (the wrentit, Nuttall's white-crowned sparrow, and song sparrow) for the seven years 1981–87.

"By the time the 1987 breeding season rolled around, the decrease in young birds [those hatched in 1986] was even greater than the decrease we had detected in the summer of 1986," DeSante said, "suggesting that still more of those birds died later that summer, or at an accelerated rate that winter. And if they did die at an extra rate that winter, isn't that awfully strange. Because those birds that were one- or two- or three-years old survived at a much greater rate. Again, the only birds that survived at a much lower rate in 1986–87 were the young and old."[31]

DeSante suggested that these puzzling results may all agree with the hypothesis that radiation from Chernobyl was the culprit. He said, "I believe that if low-level radiation is working through the immune system, it would preferentially affect the very young, whose immune systems are just developing, and the old, whose immune systems are breaking down. And that might be the reason for what we saw in 1986."[32]

Although the numbers of birds netted in the summer of 1988 at Palomarin, were an improvement over 1986 and 1987, DeSante does not expect them to achieve their pre-1986 levels until 1990 or 1991.[33]

Ornithologists are generally in agreement that birds can be regarded as early warning systems for man because they are extremely sensitive to the environment—like the canary in the coal mine. The miner never knew when poisonous gases were accumulating to dangerous levels. When the canary died, the miner hastened to get out. Did birds send a similar message to humanity in the summer of 1986, this time about the dangers of low-level radiation, particularly to especially sensitive members of the human race, such as infants and ailing adults?

CHAPTER 4

DISASTERS AT SAVANNAH RIVER

On October 1, 1988, news of nuclear accidents at the government's Savannah River nuclear weapons plant in South Carolina made headlines across the country. The accidents were described as "among the most severe ever documented," yet they were kept secret for over 20 years.[34] From information revealed in hearings conducted by Senator John Glenn, two nuclear rod "meltdowns" occurred in November and December of 1970, which could give rise to a release of large amounts of radiation. If such radiation escaped from the containment structure, the consequences to public health could have been comparable to the Three Mile Island accident of 1979. Within weeks of the disclosure, two other nuclear weapons plants were closed for safety reasons. Public suspicion that there was something seriously amiss was met by official assurances that the plants had been "operating with an adequate margin of safety."[35] E. I. du Pont de Nemours & Company, which operated the Savannah River Plant at the time of the accidents, ran a full-page advertisement in *The New York Times* that claimed "the radioactivity given off was kept within the building" and "no one, on or off the site, was ever harmed."[36]

The Savannah River nuclear weapons plant plays a prominent role in the history of nuclear technology. Its reactors manufacture tritium, a key element for modern thermonuclear weapons. After more than 30 years of operations, it is now one of the most radioactive places on earth. Almost a billion curies of high-level nuclear wastes are stored there, comprising more than half of the U.S. government's inventory.[37]

News of the Savannah River accidents raised the question of whether there were statistically significant indications of adverse health impacts in South Carolina and nearby states. Results of an examination of government databases were startling: after the two accidents in November and December, 1970, radioactivity had increased significantly in the milk and rain of South Carolina and throughout the Southeast.[38] Peaks in infant and total mortality showed up immediately following the accidents, and disturbing longer-term mortality trends appeared also in the region.

Radioactivity in South Carolina's rain, as measured for December 1970, jumped six-fold over the same month in the previous year (see Figure 4-1).[39] This jump in beta radiation occurred immediately after the accidents in November–December 1970. The rise was significantly above the local trend in the preceding 22 months and it was three times greater than the U.S. rise.[40] Also, in the Southeast as a whole, radioactivity in the rain doubled over the previous December (1.2 times greater than the U.S. rise). The average reading for the Southeast was higher than any other region in the country: five times higher than the Northeast and West and 70 times higher than the Midwest.[41]

The December 1970 readings of gross beta radioactivity in the air (in contrast to the rain) also showed a pocket of abnormally high activity in the Southeast states—with North Carolina, South Carolina, and Alabama having the highest readings east of the Mississippi.[42]

Milk was also contaminated. Radiation readings indicate that the level of strontium-90 found in South Carolina's pasteurized milk during the summer following the Savannah River Plant accidents rose significantly over the previous summer. Whereas, the level declined in the milk in the rest of the country (see Figure 4-2).[43]

The annual July readings were used here, as is the convention, because by then the dairy cows have ingested and transferred to

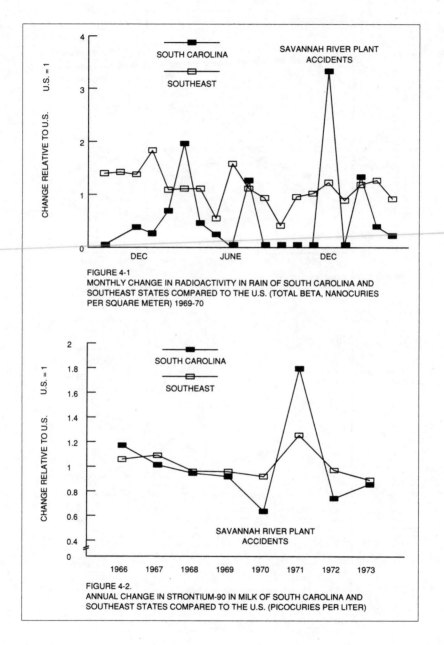

FIGURE 4-1
MONTHLY CHANGE IN RADIOACTIVITY IN RAIN OF SOUTH CAROLINA AND
SOUTHEAST STATES COMPARED TO THE U.S. (TOTAL BETA, NANOCURIES
PER SQUARE METER) 1969-70

FIGURE 4-2.
ANNUAL CHANGE IN STRONTIUM-90 IN MILK OF SOUTH CAROLINA AND
SOUTHEAST STATES COMPARED TO THE U.S. (PICOCURIES PER LITER)

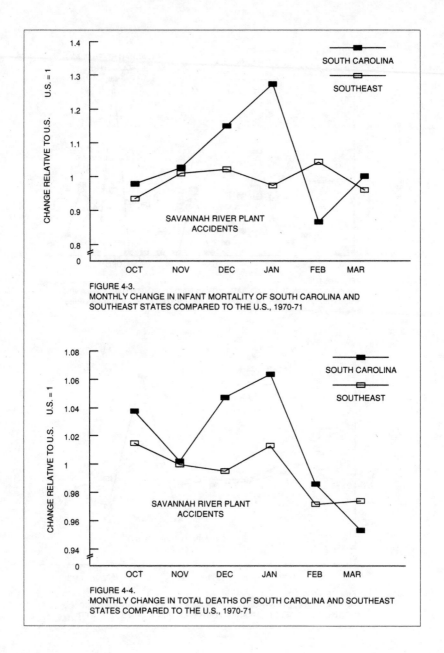

FIGURE 4-3.
MONTHLY CHANGE IN INFANT MORTALITY OF SOUTH CAROLINA AND
SOUTHEAST STATES COMPARED TO THE U.S., 1970-71

FIGURE 4-4.
MONTHLY CHANGE IN TOTAL DEATHS OF SOUTH CAROLINA AND SOUTHEAST
STATES COMPARED TO THE U.S., 1970-71

their milk the strontium-90 deposited onto pastureland by the winter's precipitation. At the same time, the summertime levels of strontium-90 in milk produced just 25 miles northeast of the plant were among the highest recorded anywhere in the country, almost twice the U.S. average.[44]

Immediately after the elevated radioactivity was found in the rain, South Carolina's infant mortality rate in January 1971 peaked at 24 percent above the previous January (see Figure 4-3). In contrast, it declined in the U.S. and Southeast during the same period.[45]

Total deaths in South Carolina also diverged significantly from the rest of the country during the months immediately following the accidents, declining six percent slower than the U.S. since the previous January (see Figure 4-4).[46]

A second, more significant and protracted peak in infant mortality occurred in South Carolina in the summer of 1971. It rose by 15 percent during the five months (May–September) over the previous summer. During the same period it declined in the rest of the Southeast and in the U.S.[47] Babies born in those months would have been in their first and second trimesters at the time of the accidents, when their pregnant mothers might have been exposed to elevated levels of radiation in the food and environment, and by the summer would have been exposed to high levels of strontium-90 in the milk.

A third and even more significant peak occurred three years after the accidents. South Carolina's annual infant mortality rate peaked in 1973, after rising two years in a row. On the other hand, rates were declining in the rest of the Southeast and in the U.S. (see Figure 4-5).[48]

Similarly, South Carolina's annual total mortality also peaked in 1973 (see Figure 4-6). From 1968–73, its total mortality increased by over three percent while U.S. mortality declined by more than three percent.[49]

Even more disturbing is the fact that the three-percent increase in South Carolina's mortality during 1971–73 was the locus of a

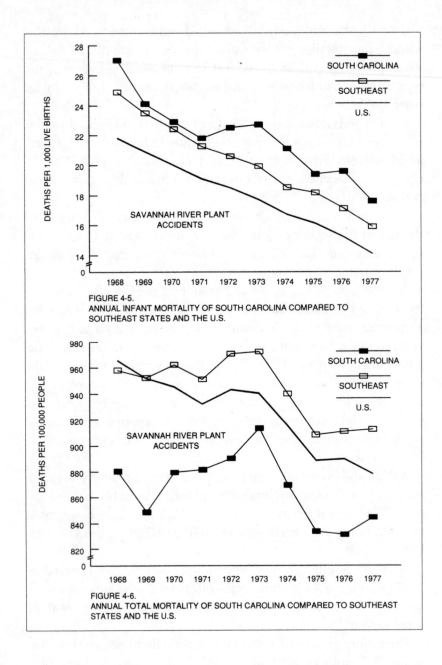

FIGURE 4-5.
ANNUAL INFANT MORTALITY OF SOUTH CAROLINA COMPARED TO
SOUTHEAST STATES AND THE U.S.

FIGURE 4-6.
ANNUAL TOTAL MORTALITY OF SOUTH CAROLINA COMPARED TO SOUTHEAST
STATES AND THE U.S.

two-percent peak throughout the Southeast, which, in turn, was the center of a statistically-significant peak (one percent) in the national rate.[50] During 1972 and 1973, more than 130,000 deaths occurred in the U.S. in excess of the number expected for those years.[51] As shown in Figure 4-6, the Southeast—and especially South Carolina—was the center of this significant excess.

In all, more than 20,000 deaths in South Carolina exceeded normal U.S. death rates throughout the 1970s and early 1980s.[52] After the 1973 peak, the excess risks of mortality from all diseases declined during the successive five-year periods 1974–78 and 1979–83.[53]

During the 15 year period, 1968–83 (starting two years before the 1970 accidents to 13 years afterwards, and excluding 1972, for which the government did not provide complete figures), the death rate from infant diseases in South Carolina continued to diverge from the rest of the country, increasing 13 percent faster than the U.S. average.[54] Infant mortality from birth defects showed an even more startling increase, rising 25 percent faster than in the U.S. and resulting in a greater than 2,400-fold leap in excess deaths (see Table 4-1).[55]

South Carolina also experienced a three-fold increase in excess lung cancer during 1968–1983. These excess deaths occurred primarily in counties surrounding and downwind of the reactors, as illustrated by the map in Figure 4-7. The bulk of the excess deaths from lung cancer took place about ten years after the reactor accidents. In fact, at the time of the accidents, South Carolina's lung cancer death rates were about three percent lower than the rest of the country. By 1979–83, however, they were about three percent higher than the rest of the U.S. (see Table 4-1).

These rates of excess lung cancer death post life-time risks that are more than 1,000 times higher than some EPA standards for increased cancer risks. EPA banned pesticide residues in food that cause one extra case of cancer per million people over a life time. Estimated excess lung cancer risks in South Carolina, in comparison, totalled over 1,700 deaths per million people. Another EPA

TABLE 4–1.

STANDARDIZED MORTALITY RATIOS AND
SIGNIFICANT EXCESS DEATHS
IN SOUTH CAROLINA, 1968–83

	CAUSE OF DEATH			
STATISTIC	ALL DISEASES	LUNG CANCERS	INFANT DISEASES	INFANT BIRTH DEFECTS
STANDARDIZED MORTALITY RATIOS (SMRS)				
1968–83 change	–3%	6%	9%	24%
1968–73 (ex 1972)	1.10	0.97	1.03	0.95
1979–83	1.07	1.03	1.12	1.18
SIGNIFICANT EXCESS DEATHS				
1968–83 total	24,129	779	697	154
1968–83 change	–23%	357%	56%*	2,438%
1968–73 (ex 1972)	8,391	88	0	8
1979–83	6,469	402	299	203

* This increase is for 1974–83, since there were no significant excess deaths from infant diseases during 1968–73 (a change cannot be computed from zero).

STANDARDIZED MORTALITY RATIOS (SMRs) are the deaths from a certain cause observed in a certain race-sex-age cohort, divided by the number of deaths expected with corresponding U.S. rates. A value of 1.00 means deaths are occurring at the expected U.S. rate; a value below 1.00 means observed mortality is better than the U.S. norm, and above 1.00 means it is worse.

SIGNIFICANT EXCESS DEATHS include the number of observed deaths that are significantly higher than expected for each race-sex cohort compared to corresponding age-specific U.S. means. Deaths in the estimate include only those for which there is greater than 99% confidence that they are not due to chance.

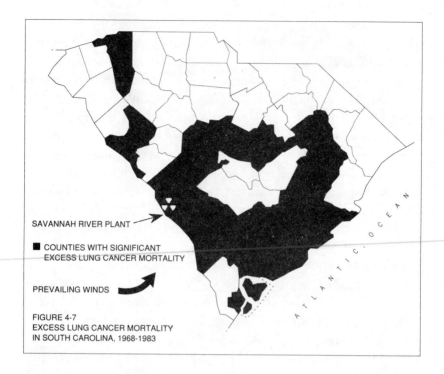

SAVANNAH RIVER PLANT

■ COUNTIES WITH SIGNIFICANT
EXCESS LUNG CANCER MORTALITY

PREVAILING WINDS

FIGURE 4-7
EXCESS LUNG CANCER MORTALITY
IN SOUTH CAROLINA, 1968-1983

standard, for exposure to uranium mill tailings, allows up to one extra case of lung cancer per 100,000 persons. The excess risks of lung cancer death in South Carolina are more than one hundred times greater than even this less stringent limit.

The pattern of excess death described here is corroborated by little-known cancer maps published by the EPA in 1987.[56] These revealed that during the 1970s, increases in white male lung cancer in counties surrounding the Savannah River Plant were among the most dramatic in the country.

South Carolina's nonwhite male cancer mortality rates rose 35 percent faster than figures for the rest of the country. In the two states surrounding South Carolina (Georgia and North Carolina) nonwhite male cancer mortality increased 28 percent faster than the rest of the U.S. In the next four surrounding states (Alabama, Florida, Tennessee, and Virginia) it increased 13 percent faster than the U.S. And in the remaining five states in the Southeast (Delaware, Kentucky, Maryland, Mississippi, and West Virginia, and the District of Columbia), it increased at the same rate as the rest of the country. The further a nonwhite male lived from South Carolina, the less likely he was to die of cancer.[57]

Significant peaks in radiation and mortality in the South may have been linked by the ingestion of elevated levels of radioactive contaminants in food and drink. The sharp jump depicted in figure 4-2 of strontium-90 in South Carolina's summer milk supply indicates widespread contamination of foodstuffs after the accidents. The unusually high readings of strontium-90 in milk produced within 25 miles of the plant were topped by even higher levels later in 1971, which then fell sharply in the following year.[58] A similar strontium-90 peak, late in 1971, was found in public drinking water supplies fed by surface water sources such as rivers and streams. The heightened concentration of strontium-90 in the autumn is consistent with a large atmospheric release of radioactivity the previous winter. It can take several months for contaminants deposited by precipitation to run off into rivers and streams. As with the milk readings, the average radiation level in public drinking water supplies dropped by over eighty percent in 1972.[59]

By far the greatest concentration of strontium-90 was found in local produce and game—fish from the Savannah River had two thousand times the strontium concentrations in water supplies. Strontium-90 accumulates in fish especially when calcium in the water is low, affecting fresh-water fish more than saltwater fish. High strontium-90 readings were also found in grains, vegetables, fruit and poultry grown within 25 miles of the plant.

The health effects of strontium-90 contamination from the plant may have been widespread. Fruit and other agricultural products grown in the Southeast are shipped throughout the country: for example, oranges from Florida and peaches from Georgia. The effect of the 1970 accident can be detected in fruit and vegetables consumed as far away as New York City. In 1971, fruit on sale in

New York City markets contained strontium-90 levels that were seven times that of its milk, which was not shipped from the Southeast. Moreover, the 30 year half-life of strontium-90 means it will contaminate soils, root vegetables, and deep-rooted fruit trees for decades.[60]

To illustrate the high level of contamination after the accidents, Table 4-2 compares the maximum 1971 strontium-90 readings within 25 miles of the plant with average readings for New York City food in 1982. The levels found in catfish and bream caught in the Savannah River were more than 100,000 times higher than the average for fresh fish in New York City (much of which was from the ocean). The concentration in collard greens grown near the plant was more than 50 times higher than the level found in vegetables in New York City; grains were eleven to 40 times more contaminated, poultry was 33 times higher, and strontium-90 in milk was eight times higher near the Savannah River Plant than in New York City. (Eleven years of decay and differences in soil uptake may account for a fraction of South Carolina's higher readings, but it is unlikely that such factors could explain these massive discrepancies.)

An adult who ate a quarter pound of catfish from the Savannah River during 1971 would have consumed more than five times the official daily radiation limit set a decade earlier.[61] An infant that ate the same amount would have received a radiation dose equivalent to about twenty chest X-rays.[62] Moreover, a protracted exposure to ingested beta emitters may be one thousand times more harmful to cell-membranes than a brief external exposure to X-rays, as de-tailed in the Methodological Appendix.

After the Savannah River accidents, readings of strontium-90 in the bones of young children in South Carolina rose by forty-five percent, while they fell by eleven percent in the Southeast as a whole and by nineteen percent in the Northeast. Throughout 1971, South Carolina had the highest readings of strontium-90 in human bones recorded anywhere in the country, and the average for children under ten was more than twice that of the Northeast. After

TABLE 4-2.

STRONTIUM-90 IN FOOD
AROUND THE SAVANNAH RIVER PLANT IN 1971
COMPARED TO NEW YORK CITY IN 1982

FOOD TYPE (OR PROXY)	(A) MAXIMUM 1971 CONCENTRATION AROUND SRP[a]	(A) RELATIVE TO AVERAGE 1982 CONCENTRATION IN NEW YORK CITY
FISH	23,000[b]	115,000[d]
VEGETATION	75,000[c]	8,523[e]
PLUMS	160	62[f]
COLLARDS	500	57[e]
OAT, RYE, WHEAT	250	40[g]
CHICKEN	10	33[h]
CORN	70	11[g]
MILK	26	8[i]
DRINKING WATER	10[j]	na

[a] Readings were taken at various places within a 25-mile radius of the plant. Concentrations are expressed in picocuries per kilogram (pCi/kg) for solids and picocuries per liter (pCi/l) for liquids. One liter of water weighs a kilogram.

[b] Includes strontium-89 as well as strontium-90.

[c] Includes all nonvolatile beta-emitters, such as strontium-90.

[d] Relative to 0.2 pCi/kg for fresh fish in New York City.

[e] Relative to 8.8 pCi/kg for fresh vegetables in New York City.

[f] Relative to 2.6 pCi/kg for fresh fruit in New York City.

[g] Relative to 6.2 pCi/kg for whole grain in New York City.

[h] Relative to 0.3 pCi/kg for poultry in New York City.

[i] Relative to 3.2 pCi/kg for dairy products in New York City.

[j] Includes only public drinking water supplies that use surface water (i.e., rivers, streams, lakes) as their source.

na = not available.

these high 1971 readings, the government stopped publishing data on strontium-90 in human bones.[63]

The peak in infant mortality observed three years after the accidents may have been caused by accumulated strontium-90 in the bones of prospective mothers, which can be particularly harmful to unborn babies. Concentrations of strontium-90 in the bone normally peak around three years after ingesting contaminated food. Damaged immune systems increase the risks of infection and can make a pregnant woman reject her fetus as a foreign body. As a result, the risks of miscarriages, premature births, low birth-weight babies, and infant mortality can increase dramatically.[64] Similar three-year latency periods were observed for premature births and miscarriages after major incidents of radioactive fallout from atmospheric weapons tests.[65]

A three-year accumulation of strontium-90 in the body could also explain the delayed peak in total mortality, which primarily involved deaths from heart diseases, as well as from cancers and other causes. The theory is that the free radicals, produced by radiation sources within the body, oxidize the low-density cholesterol and cause it to become more readily deposited in arteries, thus blocking the flow of blood and inducing heart attack.[66] An extraordinary peak in South Carolina's nonwhite mortality from cardiovascular diseases showed up during the early 1970s, compared to a similar but less dramatic peak among whites. Differences in diet, socio-economic level and medical care may have contributed to a differential effect from ingested low-level radiation.

Could something other than the Savannah River Plant accidents explain the unusual mortality and radiation phenomena? Could the high death rates have been due to chance? The probability is less than two and a half percent that total mortality peaks in the month immediately following the accidents were the result of random variations above the U.S. trend. The probability is less than half a percent that the peaks in mortality six months later were due to chance. And the probability is less than one in a million that mortality peaks three years after the accidents were a chance result.

But the question remains, if the accidents did not cause these significant mortality increases, what else could have been responsible?

Evaluation of different factors, other than the nuclear plant accidents, did not reveal any obvious alternative explanations for the excess deaths. First of all, adjustments were made for differences in age, sex, and racial distributions in the population, so these factors cannot account for the excess deaths. In addition, over 40 measures of industrial toxics, pesticides, urban pollution, smoking, and socio-economics were considered as possible alternative explanations. There was no evidence that smoking, poverty or pesticides rose dramatically at the time of the accidents and then declined, as would have had to have been the case to explain the mortality peaks in the early 1970s. None of the environmental, behavioral or biological factors were found to be both geographically associated with the excess deaths and temporally associated with their occurrence.

To test the significance of geographic divergences in other forms of pollution, South Carolina was compared with the rest of the country, and counties within the state that were considered at higher risk of exposure to fallout from the plant were compared with those at lower risk. The counties were divided into two groups: the seventeen most northwestern counties comprised the group hypothesized to be at lower risk, and the 29 coastal plain counties made up the hypothesized higher risk group.[67] It was found that the higher-risk counties experienced greater radioactivity increases and higher rates of excess mortality than the lower-risk counties. However, none of the environmental factors passed *both* tests of significant geographic divergence, that is, none were significantly higher in South Carolina than in the U.S. and in the higher-risk counties than in the lower-risk group.[68]

Can the rate of cigarette smoking be considered a factor? First, smoking does not explain why the excess deaths from lung cancer were observed primarily in counties downwind of the Savannah River Plant. Second, per-capita consumption of cigarettes has been less in South Carolina than in the rest of the country since World War II. South Carolina's low lung cancer mortality at the beginning of the 1970s (three percent below the national average) is consistent with its lower smoking rate during the 1950s and 1960s. Smoking was increasing more quickly in South Carolina than in the rest of the country, but by 1979 it was still four percent less than the national average and eight percent less than the Southeast average. So smoking alone cannot explain why lung cancer mortality was higher in South Carolina than in the U.S. during the early 1980s.[69]

It is also unlikely that the elevated lung cancer rates were due to occupational exposures to asbestos and other materials that were used in the ship building and repair industries along South Carolina's Atlantic Coast, especially during World War II.[70] During the 1960s and in the 1970s, lung cancer rates in many coastal counties improved relative to the U.S., along with a decline in asbestos usage.[71] In 1972, the U.S. Occupational Safety and Health Administration issued health regulations for asbestos, the first health regulation issued by OSHA.[72] Moreover, the worst increases in lung cancer mortality relative to the U.S. during the 1970s did not even occur in the coastal counties, but rather in inland counties adjacent to the Savannah River nuclear plant, in Georgia as well as South Carolina. The ship-building argument does not explain the excess lung cancer deaths found in many inland counties downwind from Savannah River.

Mean household income, however, passed two tests of significance for geographic divergence: households in the higher-risk counties had less income than in the lower-risk counties, and mean income for the whole state was below the national average. The question is, could lower household incomes (about $1,300 per year less than elsewhere in 1980) explain the three-fold increase in excess deaths from lung cancer and the 2,400-fold leap in excess birth defects that were observed after the accidents, especially in the counties downwind of the plant? Was there a local economic recession that bottomed-out in 1973 that could have caused the unusual peaks in mortality? Even if these conditions were satisfied, what socio-economic explanation could there be for the extraordinary monthly peaks in 1971 relative to the rest of the country? Poorer living conditions and diets would aggravate the impacts of radiation exposures, but there were no obvious socio-economic conditions that could explain the significant mortality patterns observed after the accidents.

Finally, could another source of radiation have caused the elevated readings in early 1971? There were no operating commercial nuclear reactors in the entire Southeast in 1970 and early 1971.[73] Between October 14, 1970 and November 18, 1971, China did not conduct a single atmospheric weapons test; China was the only nation that still performed such tests in the northern hemisphere at the time. Zero radiation was detected in southern Florida (Miami) throughout 1970 and 1971, indicating that South Carolina's high levels of radioactivity could not be due to bomb testing by the French government in the South Pacific. And no underground tests were carried out in Nevada during 1971 that might have leaked radioactivity into the environment until the Embudo shot on June 16, 1971.[74]

However, there was leakage from an underground test, called Baneberry, which took place in Nevada on December 18, 1970. As a result, the highest December 1970 readings of airborne radiation were recorded in Utah, with concentrations almost two hundred times higher than the national average, and seven times higher than nearby Idaho, which had the second highest reading in the country.[75] One would expect these states to have the highest readings because they are adjacent to Nevada, due north-northeast of the test site, in direct line of the prevailing winds. Also, as expected, maximum readings diminished with distance downwind from the site, with the lowest readings in the Northeast of the U.S.

The question is, could the unusual radioactivity readings in North and South Carolina, which were the highest on the East Coast, have been caused by Baneberry? North Carolina's reading was the highest in the country after Utah and Idaho, almost 15 times higher than the average for the Southeast. Some kind of magic vacuum, downwind of Savannah River, would have been needed to draw the Baneberry radiation from Nevada, over 2,000 miles away, without leaving a trace en route. Unlike local "hotspots" due to precipitation, air concentrations cannot spontaneously "re-concentrate" to give a localized concentration that was much higher than the surrounding region without violating fundamental laws of thermodynamics.

Let us again look at the strontium-90 content in the milk. Even though the prevailing winds blow to the northeast, strontium-90 in the summer milk in the Northeast states fell by thirteen percent compared to the previous summer. Strontium-90 readings in North and South Carolina's summer milk, in contrast, rose by fifty percent. In the rest of the Southeast it rose by only two percent.[76] Moreover, strontium-89 was found in South Carolina's milk in December 1970, one of only four places in the country that had reportable levels that month.[77] Since strontium-89 has a half-life of only fifty days (unlike the long-lived strontium-90), this finding suggests that the recorded contamination in foodstuffs was from a recent localized event, and not from worldwide bomb fallout or old fission products stored at Savannah River.

This analysis presents suggestive evidence that radiation was released from the Savannah River Plant during the November-December 1970 accidents. By itself, it cannot prove that anyone died from such a release, but without a plausible rival hypothesis, the trail of evidence presented here—including peaks in environmental radiation, radiation in food, radiation in human bones, and peaks in mortality—certainly points in that direction.

CHAPTER 5

THREE MILE
ISLAND

With news bulletins announcing a general emergency at the Three Mile Island (TMI) nuclear reactor, the morning of Wednesday, March 28, 1979 marked the beginning of a traumatic event in American history. In a crowded press conference in Harrisburg, Pennsylvania the day after the accident began, Ernest Sternglass urged pregnant women and young children to leave the area. In *Secret Fallout*, his account of the accident, Dr. Sternglass recounted his harrowing approach into the Harrisburg airport when his survey meter indicated radiation readings were 15 times above normal. Even within the room at the State Capitol where the news conference was being held, the readings were three to four times above normal. "I felt acutely," wrote Dr. Sternglass, "the great difficulty of having to explain, without causing a panic, the seriousness of the situation that already existed for the pregnant women and children."[78]

Two days later, Governor Richard Thornburgh ordered their evacuation, though much of the damage had already been done. Transcripts of the Nuclear Regulatory Commission's agonized deliberations during those first three days indicate the great confusion that caused the delay. Chairman Joseph H. Hendrie was recorded saying:

> It seems to me that I have got to call the Governor . . . to do it [evacuate] immediately. We are operating almost

*totally in the blind, his information is ambiguous, mine
is nonexistent, and—I don't know, its like a couple of
blind men staggering around making decisions.*[79]

Another revealing comment came from Roger Mattson, Director of
the Division of Systems Safety:

*I'm not sure why you are not moving people. Got to say
it. I have been saying it down here. I don't know what we
are protecting at this point. I think we ought to be
moving people.*[80]

According to radiation records submitted to the Kemeny Commission, convened by President Carter to evaluate the impact of the accident, most of the radiation had been released before the evacuation was ordered. The environmental consultants retained by the Metropolitan Edison Company, the owner of the damaged reactor, reported:

*Based on techniques used in this analysis, dose estimates
are consistent with the release of seven million curies of
noble gases in the first one and one half days of the
accident, two million in the next three days and one
million in the next three days, and a relatively small
amount thereafter.*[81]

Thus by the time the evacuation was ordered on the third day after the accident began, the bulk of the estimated fourteen curies of iodine-131 had already been released. The Kemeny Commission's report concluded that most of the resulting thyroid exposure was due to inhalation, rather than ingestion of drinking water and milk. Therefore most of the damage to the developing thyroid of the fetus had occurred by the time pregnant women began to leave.

The report also revealed that during the period of highest releases, from 10 a.m. on Wednesday, March 28th to 7 a.m. on Thursday,

March 29th, the winds were blowing to the north, northwest, and west at six to nine miles per hour, sending the radioactive gas toward upstate New York and western Pennsylvania (see Figure 5-1).

The Kemeny Commission concluded its summertime investigation without considering any data on the health effects of the windborne radiation. The vital statistics section of the Pennsylvania Department of Health, under the direction of Dr. George Tokuhata, refused at that time to divulge any current information on infant mortality in the immediate area around Harrisburg in the summer of 1979 on the grounds that it would take many months to process and review the data.

Though they have never been the subject of an official inquiry, the monthly mortality data eventually published by the federal government document the lethal consequences of the TMI radiation releases. An examination of the monthly changes in infant and total mortality in Pennsylvania and nearby areas as originally published in the government's *Monthly Vital Statistics Report* indicates statistically significant rises did in fact occur shortly after the accident. The National Center for Health Statistics took about four months to publish the state-by-state data in the monthly reports. The full impact of the radiation was not recorded until May, June and July of 1979, because of normal late filings of birth and death certificates. Reports for these months were published in October and November of 1979, by which time the Kemeny Commission had already submitted its final report.

The hypothesis that these abnormal mortality rises were associated with radioactive releases from TMI is strongly supported by the following considerations: First, large amounts of iodine-131 and other fission gases were released from the plant in the first two days before the order to evacuate pregnant women and children was issued. Second, infant mortality peaked three or four months after the initial release took place. This corresponds to the period when highly active fetal thyroids, which control growth hormones, would have taken up the radioactive iodine-131, and thus could explain the large increase in immature and underweight babies who died

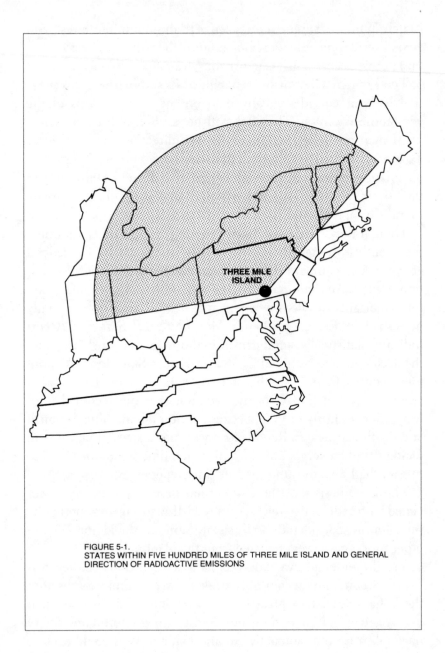

FIGURE 5-1.
STATES WITHIN FIVE HUNDRED MILES OF THREE MILE ISLAND AND GENERAL
DIRECTION OF RADIOACTIVE EMISSIONS

of respiratory distress, as reported by hospital records. Third, the greatest infant mortality increases took place in areas closest to the plant, diminishing with the distance away from Harrisburg and Pennsylvania. In sharp contrast, states to the west and south experienced declines in infant mortality rates.

Pennsylvania's infant mortality rate moved from well below the U.S. average before the accident to the highest rate in any state east of the Mississippi. Mortality in upstate New York, north-northeast of TMI, was also seriously affected, because it was downwind of the radioactive emissions. Mortality also increased in Maryland and the District of Columbia, perhaps because the Susquehanna River carried the runoff of fission products to the south. Milk produced in counties around TMI served cities throughout the Mid-Atlantic area and was another potential source of exposure. As with infant mortality, significant rises in total mortality also took place in the same areas at the time.

From the three months of 1979 prior to the accident to the four subsequent months, the infant mortality rate rose in Pennsylvania by almost sixteen percent, in Maryland by 41 percent. In comparison, New York City, well to the east of Harrisburg, declined by more than six percent, and U.S. infant mortality declined by almost fifteen percent during the same period.[82]

Infant deaths in Pennsylvania rose from 141 in March of 1979, just before the accident, to a peak of 271 in July, declining again to 119 by August. The highly significant rise occurred in the summer months when infant mortality is normally at its lowest levels. There were 242 more infant deaths than expected in Pennsylvania during the four-month period following the accident, and there were corresponding excesses in upstate New York and in Maryland.

The divergence in the tri-state infant mortality from the rest of the country was based on 114,750 birth certificates and 1,651 infant death certificates. The numbers involved were so large that the probability that this divergence resulted from chance is one out of trillions.[83]

A similar adverse pattern was observed for total mortality in the three states. The monthly average number of deaths in the summer

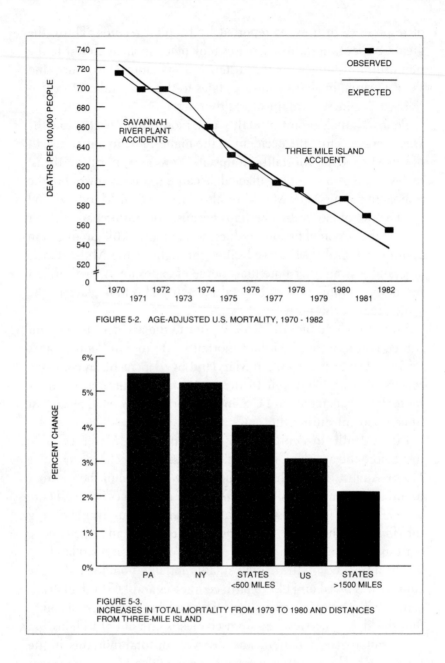

FIGURE 5-2.   AGE-ADJUSTED U.S. MORTALITY, 1970 - 1982

FIGURE 5-3.
INCREASES IN TOTAL MORTALITY FROM 1979 TO 1980 AND DISTANCES
FROM THREE-MILE ISLAND

of 1979 increased by two percent over the average for the three months prior to the accident in Pennsylvania, Maryland and up-state New York, while declining by five percent in the rest of the U.S.[84] Again the divergence was too great to attribute to chance.

Significant mortality increases continued in the years following 1979, perhaps due to latent effects on further releases from the venting of the damaged reactor in 1980. In Figure 5-2, the observed 1970–82 U.S. age-adjusted mortality rates are compared with those expected from the average annual decline of two and a half percent that occurred prior to the TMI accident. The observed jump in U.S. mortality from 1979 to 1980 was as significant as the divergence during 1970–73, which may have been due to the Savannah River accidents as explained in the last chapter. The difference between the observed and expected rates during the years 1980–82 suggests that more than 50,000 excess deaths occurred.[85]

Figure 5-3 shows that the 1980 mortality increases were broadly correlated with distance from TMI. Increases in crude mortality rates in Pennsylvania and New York, the two states most directly downwind from the accident, were by far the most significant in the country. More generally, the crude mortality rate of states within five hundred miles of Three Mile Island rose almost twice as fast as the mortality rate of states more than 1,500 miles away, again, a highly significant difference.[86] Figure 5-4 shows that the states within five hundred miles of Three Mile Island accounted for two-thirds of all estimated excess deaths in the U.S. in 1980, even though these states accounted for less than half of the 1980 population. The Pennsylvania and New York share of excess deaths was roughly twice their share of the total population.[87]

Towards the close of 1979, after *Monthly Vital Statistics Report* revealed these sharp increases in Pennsylvania's, Maryland's, and New York's summertime infant mortality, public attention was anxiously focussed on comparable data for Harrisburg and Dauphin County. Dr. George Tokuhata, who was in charge of the collection and analysis of county and city vital statistics for Pennsylvania, stated that such data would not be ready for many

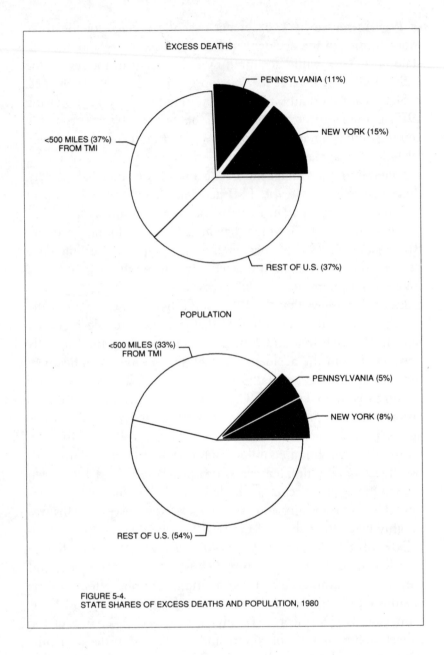

EXCESS DEATHS

PENNSYLVANIA (11%)

<500 MILES (37%)
FROM TMI

NEW YORK (15%)

REST OF U.S. (37%)

POPULATION

<500 MILES (33%)
FROM TMI

PENNSYLVANIA (5%)

NEW YORK (8%)

REST OF U.S. (54%)

FIGURE 5-4.
STATE SHARES OF EXCESS DEATHS AND POPULATION, 1980

months. He had begun a study of pregnancy outcomes for women in Harrisburg who conceived between March 1979 and March 1980, but such results were not to be available until 1982. Tokuhata was publicly criticized in the summer of 1979 by Dr. Gordon MacLeod, Pennsylvania's Secretary of Health, for withholding local monthly infant mortality data. "Such data," MacLeod said, "warrant complete candor and disclosure, not delay and denial."[88]

A decade later, the controversy over the true number of infant deaths near the reactor in the months after the accident still rages, because it lies at the very heart of some 2,500 lawsuits filed against the Metropolitan Edison Company, the owner-operator of TMI, by plaintiffs living close by, who claim to suffer from a host of radiation-induced illnesses, including: "birth defects, still births, spontaneous abortions, sterility, cancers, leukemia, hair loss, bizarre sores that won't heal, heart failure, emphysema, stroke, cerebral palsy, hypothyroidism, and a wide range of other diseases that have stricken them, their children, their farm animals, and even the foliage around them."[89]

Informal local health surveys based on laborious door-to-door canvassing describe vivid details of suspected damage to human health. Jane Lee, for example, lives on a dairy farm within sight of the reactor and has painstakingly prepared a series of overlay maps with multicolor dots representing the different afflictions and locations of the 2,500 plaintiffs near the stricken reactor. Her comments convey the suspicions and distress that continue among local residents:

> *You don't have to be a great genius to see what's going on here. Wherever the worst of the radiation blew, that's where the health effects are. The people here are the human dosimeters, and are far more reliable than the doctored instruments used by Met Ed and the state. The authorities are saying that millions of curies of iodine had to be present before you get the metallic taste, and that they only emitted fourteen. But if you have people*

all over the area complaining of the same thing in areas
where the radiation came down, then there must have
been more coming out of the plant than they want us to
believe.[90]

Dr. Tokuhata, who still maintains that no one has been harmed by
TMI emissions, never did release his promised survey of localized
pregnancy outcomes. After Dr. MacLeod's public complaint about
Tokuhata's unwarranted delays in releasing local data, then-Gov-
ernor Thornburgh requested the resignation, not of Dr. Tokuhata,
but of Dr. MacLeod! His place was taken by Dr. H. Arnold Muller,
an army specialist in "battlefield casualties," but with no public
health experience.

Under Dr. Muller, the 1979 Pennsylvania vital statistics were
finally released in November of 1980, some four months late, and
revealed that the number of infant deaths in Dauphin County was
finally tabulated as 62, a 55 percent increase over the 1978 figure,
along with a small decline in the average birth weight. While
suggestive, the numbers involved were too small to be statistically
significant. Dr. Tokuhata subsequently did admit that after the
accident there was "a significant increase in the number of babies
with low birth weights," which he attributed to "excessive use of
medications by pregnant women under stress."[91]

The venting of the damaged reactor in May and June of 1980,
however, was followed by another officially reported increase in the
number of infant deaths in Dauphin County in 1980. The final,
official 1979–80 infant mortality rate for Dauphin County was
thirty-seven percent higher than the rate for the previous two years;
during the same period, the U.S. infant mortality rate dropped by
eight percent.[92] The probability of so great a divergence occurring
by chance is less than one in one thousand.

Infant deaths from birth defects registered the most significant
increase in Dauphin County after TMI's radiation releases. Dauphin County's infant mortality from birth defects increased a
staggering forty-four percent faster than the rest of the U.S. from

the five years prior to TMI's start-up (1968–73) to the five years after the accident (1979–83); over the same period, infant birth defect mortality in Pennsylvania as a whole increased by less than six percent faster than the rest of the U.S. The probability that Dauphin County's more rapid rise was due to chance is infinitesimal.[93]

Similar extraordinary increases in infant birth defect mortality were found in South Carolina counties up to thirteen years after the Savannah River Plant accident, despite the absence of the short-lived iodine-131 to damage fetal thyroids (as described in Chapter Four). It is possible that by ingesting strontium-90, women of child-bearing age may have suffered immune system damage, which, in turn, could have led to a variety of pregnancy complications and to congenital abnormalities in offspring born during succeeding years. Dr. Ernest Sternglass was the first to suggest this possible linkage after finding a peak in 1958 of 123 deaths of congenitally-defective children in Utah, downwind of the Nevada Test Site, after an average of only 75 cases annually from 1937 to 1945.[94]

The Pennsylvania Department of Health has reported "no *significant* adverse effects" from TMI in neighborhoods close to the reactor where persons may have had the greatest exposure. Yet after the accident, residents within a multi-county area surrounding TMI experienced a metallic taste characteristic of direct exposure to radioactive releases by inhalation.[95] By investigating only the immediate vicinity of the reactor, the state may have selected areas that were too small for significant results, since statistical significance is directly related to the number of deaths in a sample.

In fact, infant mortality from birth defects in the ten-county area surrounding TMI rose over 20 percent faster than in the U.S. As would be expected with decreasing dosage levels in a larger study area, the ten-county rise was less dramatic than that of Dauphin County alone; nevertheless, it was significantly greater than the rise in Pennsylvania as a whole. These ten counties also had significant divergences in mortality from all diseases, including

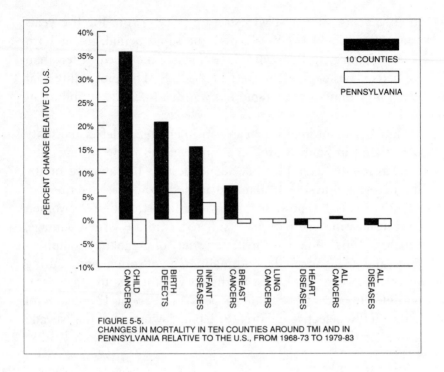

FIGURE 5-5.
CHANGES IN MORTALITY IN TEN COUNTIES AROUND TMI AND IN
PENNSYLVANIA RELATIVE TO THE U.S., FROM 1968-73 TO 1979-83

heart diseases and cancers, in particular, lung cancers, breast cancers, and especially cancers among children one to fourteen years old.[96] Figure 5-5 shows how mortality rates from each cause of death in the ten-county area either deteriorated more quickly or improved more slowly than in Pennsylvania as a whole (Table 5-1 provides details and the methodological appendix discusses possible biological explanations for such a rapid response).

Other Pennsylvania counties also experienced significant increases in mortality. At this stage we cannot disentangle quantitatively the many factors that may have influenced the geographic dispersion of the fallout and its consequences: wind and rainfall patterns, topographical barriers, runoff into the Susquehanna River, and the ingestion of contaminated milk from local dairies by persons far away. This work needs to be done. Even so, we were

TABLE 5-1.

CHANGES IN STANDARDIZED MORTALITY RATIOS
OF TEN COUNTIES AROUND THREE MILE ISLAND
COMPARED TO PENNSYLVANIA, 1968–83

Cause of Death	TEN COUNTIES AROUND TMI*			PENNSYLVANIA		
	1968–73** SMR	1979–83 SMR	Percent Change	1968–73** SMR	1979–83 SMR	Percent Change
ALL DISEASES	1.02	1.01	–1.3%	1.09	1.08	–1.5%
Heart Diseases	1.07	1.05	–1.2%	1.10	1.08	–1.8%
Infant Diseases	0.88	1.01	15.5%	1.03	1.07	3.6%
Infant Birth Defects	0.89	1.08	20.7%	1.06	1.12	5.8%
ALL CANCERS	0.99	1.00	0.5%	1.06	1.06	0.0%
Lung Cancers	0.85	0.85	0.0%	0.99	0.98	–0.8%
Breast Cancers	1.04	1.11	7.0%	1.08	1.07	–0.9%
Child Cancers	0.77	1.04	35.7%	0.97	0.92	–5.1%

* The ten counties surrounding TMI include Dauphin and Cumberland Counties and the adjacent counties of Adams, Franklin, Lancaster, Lebanon, Montour, Northumberland, Perry, and York.

** Excluding 1972, because of an incomplete government sample.

STANDARDIZED MORTALITY RATIOS (SMRs) are the deaths from a certain cause observed in a certain race-sex-age cohort, divided by the number of deaths expected with corresponding U.S. rates. A value of 1.00 means deaths are occurring at the expected U.S. rate; a value below 1.00 means observed mortality is better than the U.S. norm, and above 1.00 means it is worse. Percent changes in SMRs can thus be read as the change relative to the U.S. (e.g., mortality from infant birth defects increased more than twenty percent faster in the ten-county area than in the U.S.).

unable to find any environmental or socioeconomic factor that could account for so great a mortality divergence in Dauphin and surrounding counties other than exposure to TMI fallout.[97]

So the final, official figures of the National Center for Health Statistics would appear to contradict the claim that "no one died at TMI."

CHAPTER 6

COVER-UP

In February 1974, one month before a federal grand jury indicted seven Nixon aides for the Watergate conspiracy, an innocuous report on the Savannah River Plant appeared in an official government radiation publication, stating:

> The quantity of radioactivity released by SRP to its environs during 1971 is, for the most part, too small to be distinguished from natural background radiation and fallout from worldwide nuclear weapons tests. . . . [E]nvironmental levels continue to be far below levels considered significant from a public health viewpoint.[98]

Such explicit denials were just one of a variety of techniques used to veil the occurrence of major accidents at one of the nation's most important nuclear weapons plants.

The government has admitted it kept the Savannah River Plant accidents secret for decades. Official spokespersons said this secrecy was part of "a deeply rooted institutional practice, dating from the first days of the Manhattan Project in 1942, which regarded outside disclosure of any incident at a nuclear weapons production plant as harmful to national security."[99] In fact, the very law that established civilian control over nuclear technology with the creation of the Atomic Energy Commission (AEC) in 1947 also established a special category of "restricted data" that included all information "concerning the manufacture or utilization of atomic weapons, the production of fissionable material, or the use of fissionable materials in the production of power."[100] The disclosure of restricted data to unauthorized persons or parties was punishable by death. At the start-up of the AEC, over 80 percent of its research reports were classified as restricted data.[101]

Given this history of censorship and the centrality of Savannah River to U.S. strategic policy, it is easy to understand why there was no mention of an accident or radiation release from the plant in any published government report at the time. However, there is also evidence of a consistent and widespread effort to keep the lethal effects of low-level radiation secret, regardless of whether the radiation was released by military or civilian reactors—methods that appear to have included outright falsification of published data.

The existence of government documents that regularly published radiation and mortality data would have posed a considerable problem to any official who tried to keep such accidents and health effects secret. During the early years of atmospheric bomb testing, Congress requested independent monitoring of environmental radiation levels to act as a check on AEC operations. In 1959, President Eisenhower directed the U.S. Department of Health, Education, and Welfare (HEW) to publish monthly readings from nationwide networks that were set up to sample radiation levels in the environment. As stated in the opening paragraph of the official monthly publication *Radiation Data and Reports*, "this form of surveillance . . . forms a basis for an alerting system . . . of radioactivity in food, milk, and water."[102] If there were major releases of low-level radiation, then this "alerting system" supposedly would be able to detect them.

Another government publication serves a related early warning system function: the *Monthly Vital Statistics Report*. These state-by-state tabulations of infant mortality and total deaths can identify episodes of heightened mortality. Linking these monthly mortality and radiation data could reveal significant correlations—that is, if both reporting systems function properly.

If the data were falsified, on the other hand, then individuals from a number of federal agencies must have been involved. It is well understood that the radiation data are highly sensitive, affecting food supplies, public health, public opinion, and national security. Sitting on the editorial board of the radiation reports were not only representatives of HEW, but also reviewers from the Departments of Agriculture, Interior, Defense, and from the AEC. At the time of the Savannah River accidents, the Public Health Service of HEW published both the monthly radiation reports and the *Monthly Vital Statistics Report*. An organizational structure existed, therefore, that could alert selected individuals of unusual readings prior to publication, and hence that allowed for the possibility of supression.

By 1974, just when *Radiation Data and Reports* published their denial that any radiation was released from Savannah River Plant, the Public Health Service also began to publish significantly altered mortality data. The official mortality figures for 1971, the year following the Savannah River Plant accidents, were finally published in the bound volumes of the *Vital Statistics of the United States* in 1974. These bound volumes are the standard reference for U.S. mortality data; they are easier to use than multiple copies of the *Monthly Vital Statistics Report*, and are more widely available in public libraries. The data in the annual volumes undergo final adjustments to correct for deaths that occur in states other than where the deceased had lived (if, for example, state lines were crossed for a better hospital). These final statistics are therefore called mortality "by place of residence." The data in the *Monthly Vital Statistics Report*, on the other hand, are called mortality "by place of occurrence."

Chapter Four describes peaks in South Carolina's January 1971 mortality using data that were already corrected for errors from late filings, faulty death certificates, computer malfunctions, and other random mistakes (see Figures 4-3 and 4-4). These were published in *Monthly Vital Statistics Report* a full year after the original figures appeared.[103] Not only had the figures not changed substantially as a result of corrections made a year later, but the peaks were even more significant in the revised data than in the original reports.[104]

But in the final bound volumes published in 1974, the significant January 1971 peaks in South Carolina's infant and total mortality disappeared. Normally, such changes are marginal, but in this case, the shape of the monthly variations for both infant and total mortality in South Carolina were altered completely (see Figures 6-1 and 6-2). In the final by-residence data, South Carolina's January change over the previous year became the same or less than the U.S. change, whereas it had been significantly higher for both infant and total mortality in the revised by-occurrence data.[105]

Similar revisions also occurred after the Three Mile Island and Chernobyl accidents. The methods used to revise the vital statistics were increasingly sophisticated, however, paralleling bureaucratic improvements for handling the radiation data, as discussed below. In 1974, the same year Three Mile Island began operating and the altered South Carolina data were published, Pennsylvania and Maryland infant mortality rates were systematically revised in the *Monthly Vital Statistics Report,* twelve months after the original figures were published (see Table 6-1).

Although the Three Mile Island revisions were made earlier in the publication process than those after the Savannah River accidents, they shared the common feature that all involved alterations in the number of reported deaths. For example, South Carolina's January 1971 peak in infant mortality was eliminated by the removal of almost a third of the infant deaths that occurred in the state that month.[106] Presumably, these deaths were now missing because all of these babies had traveled from out-of-state that month to die. The subtraction of 38 dead babies in South Carolina the month after the Savannah River Plant accidents can be compared to 59 missing infant deaths in Maryland during July 1980, right after a Three Mile Island venting, and eighty-six missing in

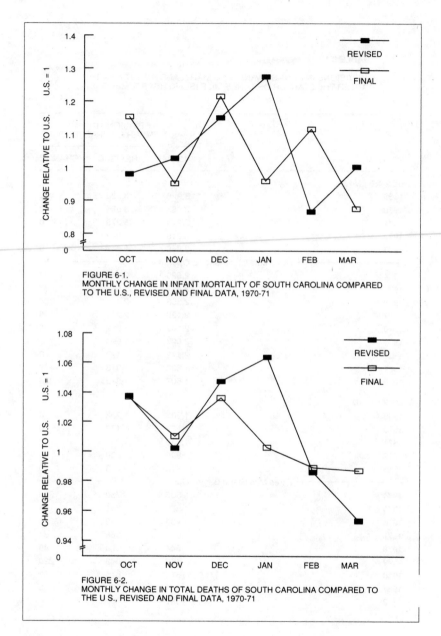

FIGURE 6-1.
MONTHLY CHANGE IN INFANT MORTALITY OF SOUTH CAROLINA COMPARED
TO THE U.S., REVISED AND FINAL DATA, 1970-71

FIGURE 6-2.
MONTHLY CHANGE IN TOTAL DEATHS OF SOUTH CAROLINA COMPARED TO
THE U S., REVISED AND FINAL DATA, 1970-71

TABLE 6-1.

REVISIONS IN PENNSYLVANIA AND MARYLAND INFANT DEATHS
AFTER THE STARTUP OF THREE MILE ISLAND (SEPTEMBER 1974)

YEAR	REVISIONS IN NUMBER OF DEATHS		
	ORIGINAL	REVISED	CHANGE
PENNSYLVANIA			
1969	3,926	3,926	0
1970	3,885	3,885	0
1971	3,278	3,278	0
1972	2,919	2,919	0
1973	2,646	2,646	0
——————————— Three Mile Island Begins Operation ———————————			
1974	2,620	2,714	94
1975	2,454	2,541	87
1976	2,353	2,411	58
1977	2,136	2,241	105
1978	2,262	2,262	0
1979	2,097	2,099	2
1980	2,179	2,179	0
1981	1,836	1,913	77
1982	1,507	1,873	366
MARYLAND			
1969	1,203	1,203	0
1970	1,177	1,177	0
1971	1,033	1,033	0
1972	806	806	0
1973	718	718	0
——————————— Three Mile Island Begins Operation ———————————			
1974	507	730	223
1975	570	761	191
1976	433	748	315
1977	594	658	64
1978	644	693	49
1979	623	843	220
1980	675	614	−61
1981	594	594	0
1982	587	587	0

Pennsylvania during July 1979, right after the Three Mile Island accident (see Figures 6-3 and 6-4). In all three cases, peaks in infant mortality were eliminated as a result of the revisions. As Table 6-1 indicates, as many as three hundred or so infant deaths a year were added or subtracted from the annual totals of Pennsylvania and Maryland reported in *Monthly Vital Statistics Report* since Three Mile Island began operating. No such dramatic changes were made in these states prior to then.

In the final data published in the *Vital Statistics of the United States,* the change in the Pennsylvania's infant mortality rate after the Three Mile Island accident was no longer significantly different from the country as a whole, largely because of the drop in the state's July infant deaths. Similarly, the change in Pennsylvania's total death was also no longer significantly higher than expected, because of an inexplicable upward revision in the number of deaths during the three months prior to the accident.[107]

Other types of revisions began to appear by 1979 that prevented the identification of significant peaks in less conspicuous ways. For every month of 1979 after the Three Mile Island accident, for example, data for California, Minnesota, and Illinois were missing from successive issues of *Monthly Vital Statistics Report.* These highly irregular omissions made it impossible to evaluate the significance of mortality increases in areas near Three Mile Island, because the baseline U.S. mortality trend could not be calculated.

Dramatic revisions were also made to the monthly data after the Chernobyl fallout reached the U.S. in 1986. Some states markedly increased their reported number of live births in *Monthly Vital Statistics Report.* Since the number of births is the denominator in an infant mortality rate, increasing this number lowers the rate. Small revisions in birth data can be expected due to late filings. As stated in every *Monthly Vital Statistics Report,* "delay in the receipt of certificates in a registration office may result in a low State count for a given month followed by a high count for the month(s) in which the delayed records are received." Thus, when corrections are made a year later, the monthly numbers rise and

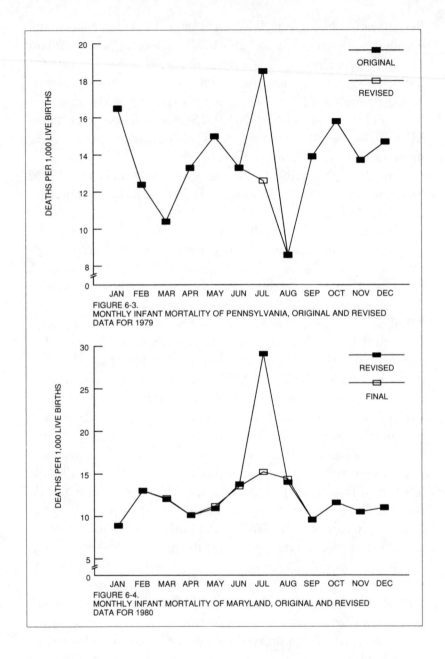

FIGURE 6-3.
MONTHLY INFANT MORTALITY OF PENNSYLVANIA, ORIGINAL AND REVISED
DATA FOR 1979

FIGURE 6-4.
MONTHLY INFANT MORTALITY OF MARYLAND, ORIGINAL AND REVISED
DATA FOR 1980

fall randomly as a result of late filings. The same can occur with late death certificates.

Revisions to the 1986 birth data for California and Massachusetts, however, were all positive and clearly non-random, adding nearly 45,000 live births to their original totals. Exactly 813 births were added for each of five successive months to Massachusetts' monthly data. The next three months, the changes were 703, 702, and 703. The nine upward revisions in California's birth data were all combinations of 5000, 4000, and 4415. At the same time, there were no major revisions in the reported number of infant deaths. The result was that Massachusett's June 1986 infant mortality rate was lowered by 76 percent and California's July infant mortality by over 25 percent. In this way, the large infant mortality peaks in the original data for California and Massachusetts after Chernobyl were eliminated.

When queried, the Massachusetts Department of Health complained of a computer error, and the California Department of Health Services said Los Angeles fell behind and filed some 30,000 certificates in one big group at the end of the year. Yet these responses do not explain the arbitrary nature of the changes made a year later, or why the major revisions just happened to begin after the Chernobyl fallout arrived in the U.S. (See Table 6-2.)

Also in 1986, unexplained changes in the publication format of *Monthly Vital Statistics Report* began to appear. Since January 1986, for example, the chart of monthly infant mortality has been missing from its customary place on the top half of the second page alongside the charts of monthly changes in births, marriages, and deaths. And since January 1987, revisions of monthly data were no longer marked, impeding their identification and analysis. Instead a generic note at the bottom of each page states, "figures for earlier years may differ from those previously published." Now, the individual changes are not only unexplained, they are also no longer identified.

Crucial omissions began years earlier in *Radiation Data and Reports*. The tables were full of omissions footnoted with a variety

TABLE 6-2.

REVISIONS IN CALIFORNIA AND MASSACHUSETTS
LIVE BIRTHS AND INFANT MORTALITY
AFTER THE ARRIVAL OF CHERNOBYL FALLOUT (MAY 1986)

| YEAR | MONTH | REVISIONS IN NUMBER OF LIVE BIRTHS | | | PERCENT CHANGE IN INFANT MORTALITY RATE |
		ORIGINAL	REVISED	CHANGE	
CALIFORNIA					
1986	March	39,826	39,826	0	0%
	April	34,675	34,675	0	0%
	Chernobyl Fallout Arrives				
	May	29,373	33,788	4,415	−13%
	June	36,591	41,006	4,415	−11%
	July	26,808	36,223	9,415	−26%
	August	32,672	37,672	5,000	−13%
	September	42,934	42,934	0	0%
	October	39,065	39,065	0	0%
	November	33,220	38,220	5,000	−13%
	December	47,471	51,886	4,415	−9%
1987	January	42,867	47,867	4,000	−9%
	February	40,946	44,946	4,000	−9%
	March	44,498	48,498	4,000	−8%
	April	35,416	35,416	0	0%
MASSACHUSETTS					
1986	March	7,282	7,282	0	0%
	April	6,999	6,999	0	0%
	Chernobyl Fallout Arrives				
	May	4,566	4,566	0	0%
	June	1,352	5,715	4,363	−76%
	July	—	5,905	5,905	na
	August	5,905	6,718	813	−12%
	September	6,340	7,153	813	−11%
	October	4,042	4,855	813	−17%
	November	9,608	10,421	813	−8%
	December	4,962	5,775	813	−14%
1987	January	6,237	6,940	703	−10%
	February	6,238	6,940	702	−10%
	March	3,366	4,069	703	−17%
	April	7,031	7,031	0	0%

of excuses: "no report received," "no sample collected," "no field estimate," or "samples were collected but no field estimates were received." As far back as 1967, the station located in Gastonia, North Carolina, 140 miles downwind of the Savannah River Plant stopped reporting radioactivity altogether. There were two exceptions to these omissions, however, revealing the irrelevance of the excuse that the station was "part of another network." The first was a reading in September 1968, just about the time when South Carolina experienced huge increases over the previous year readings. The second reading happened to be in January 1971, the month after the Savannah River Plant rod meltdown.

The rare Gastonia, North Carolina reading after the Savannah River Plant meltdown reveals another potential cover-up technique. The reported amount of radioactivity in North Carolina rain that month (January 1971) was zero. Zero readings were also extensively reported at the Columbia, South Carolina sampling station, in fact, for nine of the twelve months of 1971. These zero readings (which show up in Figure 6-5 as zero change relative to the U.S.) were reported at a time when extensive clean-up was underway, and when strontium-90 levels in milk within 25 miles of the plant were among the nation's highest. Figure 6-5 shows that zero radiation was reported in South Carolina rainfall for almost every month of the three years following the Savannah River Plant accidents, as indicated by the solid black line at the bottom of the figure. These null readings took place despite significant radiation increases in the Southeast during the same time period, and despite the probability that radiation was vented from Savannah River following the meltdown.

Closing all the sampling stations in the Southeast or having them all report "no sample" or zero radioactivity would have been too blatant, so another technique appears to have been used to divert attention from Savannah River Plant. This involved transferring high readings from South Carolina to another station in the Southeast, but located generally upwind from the plant and at least

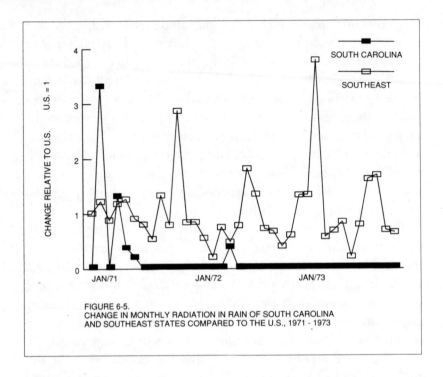

FIGURE 6-5.
CHANGE IN MONTHLY RADIATION IN RAIN OF SOUTH CAROLINA
AND SOUTHEAST STATES COMPARED TO THE U.S., 1971 - 1973

one state away. High readings at such a location would not have been easily attributable to airborne releases from Savannah River.

The station chosen appears to have been in Montgomery, Alabama, where the Atomic Energy Commission's environmental laboratory was located. The 1971 peaks in the Southeast shown in Figure 6-5 were due to extraordinarily high levels of radioactivity in the rain in Montgomery, levels similar to those recorded at the peak of large-scale atmospheric testing by the U.S. and U.S.S.R. The March 1971 readings for Montgomery were the highest in the country, thirteen times higher than the U.S. average that month, and the second highest reading for any station during any month of 1970–71. That March reading was higher than any in Montgomery during the six years prior, and higher than any there since. Even so, Montgomery continued to have the highest readings in the country for five of the nine remaining months of 1971.[108] South Carolina, on the other hand, reported zero radioactivity during all of these months.

Montgomery's enormously high values of radioactivity could have only been caused by a major release from the Savannah River Plant. No other sources of fission products could have rained-out as a "hot-spot" in Montgomery, Alabama during the spring of 1971. There were no operating reactors in the Southeast. There were no underground tests in Nevada. There were no Chinese atmospheric tests. The strategy worked at the time because no one in the scientific community or the public could have known that until June 1971 no underground or near-surface tests were conducted in Nevada for either the Plowshare or weapons programs.

By 1975, public and Congressional concern over the Atomic Energy Commission's conflicting roles as both the promoter and regulator of atomic energy grew to such a degree that the Agency was disbanded. The Nuclear Regulatory Commission was established to take over the AEC's oversight role. EPA took over HEW's responsibilities for collecting and publishing the radiation data. Publication of the monthly *Radiation Data and Reports* ceased. A far more limited and less-widely-distributed quarterly report called *Environmental Radiation Data* replaced it. This report no longer contained the detailed articles that had been published in *Radiation Data and Reports*, covering environmental radiation around government weapons facilities, radioactivity in the diet, and strontium-90 in the bone. Critical data on long-lived strontium-90 and short-lived strontium-89 in milk were reduced to a single July measurement per state.

Since its first issue in 1975, *Environmental Radiation Data* has been prepared by none other than the former AEC laboratory in Montgomery, Alabama, now called EPA's Eastern Environmental Radiation Facility (EERF). To this very same facility are now sent all the milk samples, the rain samples, and the air filters for detailed measurements of environmental radioactivity.[109] This one laboratory can now determine the measured levels of environmental radiation everywhere in the country, a set-up that could certainly simplify alteration or concealment of data.

One of the most unusual practices of EERF has been the reporting of large "negative" radiation values for milk, which are physically impossible—there is no such thing as a negative amount of radiation. According to a relatively new procedure, however, if the radiation in milk is lower than background levels that are due to cosmic rays, radon, and other natural sources, then small "negative" readings occur. Under this system, there are normally as many small positive as negative values for short-lived substances such as iodine-131 or barium-140 that decay in a matter of a few weeks. As statistically expected, the numbers are rarely larger than four or five units and average out to zero over a period of a few months to a year.

Contrary to this expectation, reports have occasionally included many more negative than positive values, with negative values of unusually large magnitudes. One such occasion was in the summer of 1982 when underground nuclear weapons tests were being conducted at the Nevada Test Site for the Reagan Administration's "Star Wars" strategic defense initiative. According to testimony at Congressional hearings, these tests were suspected of leaking significant amounts of radioactivity into the environment.[110] Since the last Chinese atmospheric bomb test was in October of 1980, and since there were no nuclear reactors in operation in the area, large positive readings of barium-140 and iodine-131 in the milk would have been an unequivocal signature of leaks from the underground tests.

Instead, barium-140 in Nevada milk reached the incredible value of negative 42 picocuries per liter in June 1982, the most significant negative reading in the nation. Out of a total of 62 barium-140 measurements reported for the U.S. that month, an astounding 57 were negative! Eight western states neighboring Nevada also had negative barium-140 measurements that diminished in magnitude with distance away from the test site. In sharp contrast, no such clear-cut pattern existed for the month of March, when the magnitude of negative radioactivity in the milk was ten times less in Nevada than in June, and there only were small positive and negative fluctuations as expected under normal conditions. (See Figure 6-6.)

Also in June 1982, large negative readings of iodine-131 were clustered in New England. That month, the Pilgrim nuclear reactor in Massachusetts had two serious radiation releases. These and earlier reported releases were the subject of a Harvard School of Public Health investigation of subsequent leukemia increases, which Senator Kennedy later used to justify a National Institute of Health study of cancer rates near nuclear reactors.[111] Figure 6-7 shows how again, as with the states surrounding the Nevada Test Site, the pattern of significant negative iodine-131 readings diminished with distance away from reported large sources of iodine-131, namely the Pilgrim reactor in eastern Massachusetts and the Millstone reactor in eastern Connecticut. In this case, reports submitted to the NRC by the operator of Pilgrim confirmed the existence of elevated radioactivity in the air near the plant that summer.[112] And as with the Nevada example, measurements in surrounding states had been near zero earlier in the year when the plant was closed.

Why have large negative readings? Why not just change the readings to near zero? The use of large negative values could be meaningful for officials who needed to know the truth. But for the public at large, these negative values would cancel the positive readings, so resulting national averages would never cause alarm.

Undoubtedly, such bizarre data manipulations have been a technique of last resort. To start with, the likelihood of a scandalous disclosure of an incriminating statistic is remote. These monthly publications are highly technical and aimed at a very limited scientific readership. Most of the scientists who read the radiation reports, for example, work for one of the federal agencies involved with atomic energy. Recognizing the sensitivity of the data for national security, such government scientists would probably report any unusual readings to proper authorities, who, in this case,

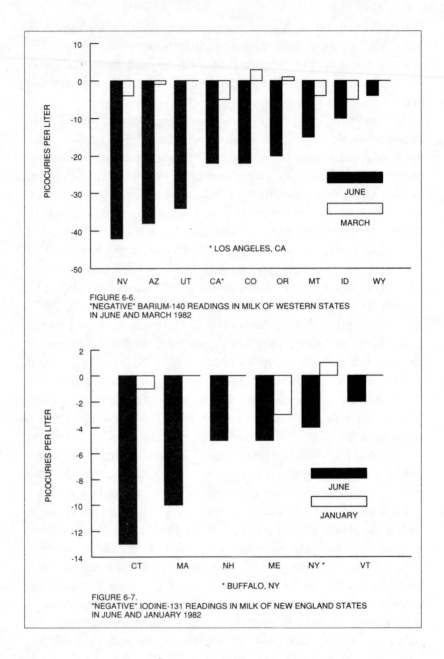

FIGURE 6-6.
"NEGATIVE" BARIUM-140 READINGS IN MILK OF WESTERN STATES
IN JUNE AND MARCH 1982

FIGURE 6-7.
"NEGATIVE" IODINE-131 READINGS IN MILK OF NEW ENGLAND STATES
IN JUNE AND JANUARY 1982

would be their government superiors. On the other hand, many scientists in the public health community who read *Monthly Vital Statistics Report* would not follow the radiation data very closely. After the atmospheric test ban treaty was signed in the early 1960s, relatively few scientists have worried about environmental levels of man-made radioactivity.

In addition, less conspicuous and problematic means were often available to suppress unusual measurements. For example, it is very difficult to interpret the tabular listings of radiation measurements, simply because the order in which the stations are listed has no relation to their geographic location. Alaska follows Alabama, South Carolina follows Rhode Island, and so on. It is necessary to computerize the data to see which readings are significantly higher than their surrounding regions.

There were no personal computers when the elimination of mortality peaks began. Conducting the type of analyses presented in this book would have been prohibitively expensive even for a large organization. This was known at the AEC, of course, because that was where computers had their first major application, and the AEC continued to be at the cutting edge of computer technology as it pursued classified weapon systems research. Even so, as recently as 1986, the EPA had not computerized the radiation data to the degree necessary for analysis.[113]

At the time of the Savannah River accidents, a decision to falsify the radiation data would have had to have been coordinated quickly. Publication of monthly data on radioactivity in the air and the rain occurred about four months after the measurements were taken. The radiation samples were collected at various stations across the country, which were operated by state health departments. Depending on the level of radioactivity found, initial readings were either telephoned or sent in writing to appropriate federal officials. It is likely that a decision to falsify data would have been difficult to implement without overcoming major bureaucratic obstacles. The establishment of EERF would have certainly simplified matters.

Changes in the data would have to be made selectively. In some cases publication of falsified data would have created greater security problems than publishing a high reading buried among the hundreds of numbers in the monthly tables. Blatant misreporting of certain radiation levels would be much more prone to whistle-blowing. It would have looked rather suspicious to radiation experts, for example, if no high readings were reported around Nevada after a highly publicized event such as the Baneberry test. So governments appeared to have sometimes used another option instead—simply to halt publication of the monthly reports altogether.

The Soviet Union, for example, ceased publication of infant mortality rates in the early 1970s. Infant mortality there rose sharply following decades of improvement, soon after its own nuclear reactors began to come on-line. The U.S. government, in contrast, could not simply eliminate the reports without raising public suspicions. Congress had ordered their publication, and by the early 1970s, there was a growing controversy over the adverse health effects of low-level radiation. At the beginning of 1970, for example, the Joint Committee on Atomic Energy held well-publicized hearings at which Drs. John Gofman and Arthur Tamplin testified that there was a direct relationship between low-level radiation and cancer in adults, fetuses and infants.

Over time, however, there were significant cutbacks in related publications and research efforts. For example, the AEC managed to terminate the research of Drs. Gofman and Tamplin at Lawrence Livermore Laboratories. In the course of examining radiation doses from the Plowshare nuclear excavation program, the scientists refused to eliminate estimates of health effects from atmospheric tests and reactor releases—their project was abruptly cancelled. The AEC also terminated the research of Dr. Thomas Mancuso at the University of Pittsburgh on worker exposure to low-level radiation at the government's Hanford reactors, after he found much larger than expected cancer rate increases. An important article by Dr. Lester Lave and co-workers at Carnegie-Mellon

University, which showed high correlations of mortality rates with strontium-90 in milk, was eliminated just prior to its publication in *Radiation Data and Reports*.[114] Dr. Alice Stewart, engaged by the Three Mile Island Fund to research the health effects of low-level radiation at government facilities, was denied access to crucial worker exposure and health data that Congress had expressly mandated for such studies.

Similar reductions in the release of key mortality data have been made by at least three state departments of health. During the period of heavy nuclear fallout in the early 1960s, Massachusetts halted the release of data on live births, infant deaths, and total deaths—but not on marriages. After revelations in 1970 that fallout from the Nevada Test Site hit the Albany-Troy area, with subsequent increases in leukemia, the New York State Department of Health curtailed the detailed annual listing of deaths by cause and location, substituting a greatly shortened summary. After large radiation releases from the Millstone nuclear reactor in 1975, the Connecticut Department of Health Services stopped publishing cancer mortality by township, which had appeared annually since the 1930s.

It is a large and difficult task to perform quality control on sizable databases such as those used to produce the monthly reports on radiation and vital statistics. Budget restrictions and the Paperwork Reduction Act have placed enormous pressures on government agencies responsible for reporting information. Nevertheless, the case presented thus far suggests nothing less than a conspiracy perpetrated against the health of the American people. This is a difficult conclusion for anyone to accept, even after the government has admitted that it lied about accidents at the Savannah River Plant for over twenty years. But it is even more difficult from the perspective of information systems experts, to come up with an explanation for the inconsistencies and revisions observed in the data that is both innocuous and plausible. Let us consider a few possibilities.

Maybe the inconsistencies and revisions are the result of random errors which had to be corrected. This defense is in the province of statistics. As pointed out in Chapters One and Two, the probability is infinitesimal that the simultaneous post-Chernobyl mortality peaks in the U.S., in Germany, and among birds occurred as a random coincidence. It is just as unlikely that the radiation peaks found in the air, rain, milk and fish, and the mortality peaks among infants and adults after the Savannah River accidents were all coincidental. And how likely is it that reporting errors just happened to occur after a major accident at Savannah River, and again after a major accident at Three Mile Island, and again after a major accident at Chernobyl, and at no other time, with each error resulting in a subsequent mortality peak that had to be revised each time? It would be absurd even to attempt to calculate such a tiny probability. In short, the revisions appear systematic.

Maybe there has been an internal policy to smooth out unusual peaks and troughs in the monthly vital statistics, a policy which had nothing to do with "national security" or nuclear accidents. Mortality is a very stable statistic, and so these fluctuations are improbable anyway. A similar policy was employed by government scientists who maintained data on atmospheric ozone. Since they were trying to develop a model of the gradual depletion of ozone in the atmosphere, they programmed their computers to ignore abnormally low seasonal dips. As a result, these top-level scientists, who were using the most sophisticated satellite data collection systems, analytic models, and computer hardware available, completely missed a massive "hole" in the ozone shield over Antarctica.

The application of such a "smoothing-out" policy to U.S. vital statistics would not only lead scientists to miss mortality peaks following nuclear accidents and other environmental exposures, but it would also be fraudulent. There is no mention of such a policy in the *Monthly Vital Statistics Report* or in the *Vital Statistics of the United States*. If this policy exists, it would defeat the very purpose of publishing monthly vital statistics as a national health alert system. The accuracy of all U.S. mortality data would be suspect.

Maybe the changes were made by state health departments attempting to minimize local political problems, without federal complicity. This would mean that California, Maryland, Massachusetts, Pennsylvania and South Carolina all committed similar falsifications over a time period that spanned at least a decade and a half. If this was such common practice across the states, could it have been kept a secret over so many years without some form of coordination?

The most disturbing explanation, but unfortunately the most convincing one, is that there has been a conspiracy orchestrated from the highest reaches of government and begun in the early days of the Cold War. A public servant would need an overwhelming reason to perpetrate such a fraud. What could be more compelling than a perception that the changes were necessary as a matter of national security—and if national security were the overriding concern, how could the federal government not be involved? On the other hand, why would anyone go to the trouble of doctoring numbers if the health effects were not real, or were not very serious? It would be difficult to find people concerned with national security in state agencies, but not impossible. It would be difficult to pass on secrets after the AEC was dismantled, but not impossible. Until the last few years, EPA was not even allowed to inspect the weapons plants for environmental problems. This recent development is a major reason why the disclosures of accidents at nuclear weapons plants have only just begun.

Despite all of these indications of a cover-up, database analyses of the sanitized data still provided significant results. The mortality increases were simply too significant to erase even with a concerted effort.

Even with all of the revisions published in *Monthly Vital Statistics Report* during 1987 and 1988, the significant Chernobyl findings described in Chapter Two still hold true.

The peaks in South Carolina's mortality during January 1971 disappeared in the final data published in the *Vital Statistics of the*

United States, but the more significant five-month peaks that summer remained.[115] Moreover, all of the even more statistically significant annual, multi-year and county mortality increases identified in Chapter Four used the final (altered) data, and were all adjusted for race, sex, and age.

Despite the revisions in Pennsylvania's infant and total mortality data in the final version, the divergent trends for the tri-state area during the four months after the Three Mile Island accident remained statistically significant in the final data.[116] Moreover, in Dauphin County, home of Three Mile Island, the 1979–80 increase in infant mortality over the previous two years remained highly significant in the final data as published in the *Vital Statistics of the United States.* Finally, as indicated in Chapter Five, the detailed county data on the observed and expected number of deaths in Pennsylvania since 1968 reveal such significant increases in mortality from so many causes of death after 1979, that any attempt to tamper with them would collapse the entire structure of the system of collecting and analyzing our vital statistics.

Thus, in the long run, the myriad revisions and data manipulations were insufficient to obscure completely the case-study findings of significant increases in mortality linked to releases of low-level radiation. But it is an unspeakable tragedy if Americans were kept ignorant of nuclear accidents and their consequences even if it was done in the name of national security. Invoking "national security" cannot justify denying people a chance to protect their children's health, or their own. Without full knowledge, the medical and public health communities cannot investigate significant causes of mortality. And no one can learn the lessons that might prevent accidents . . . at Savannah River, at Three Mile Island, at Chernobyl, and who knows where else?

The true story behind the evidence of a cover-up can only come from intensive Congressional investigations and hearings.

IMPLICATIONS

CHAPTER 7

ATMOSPHERIC BOMB TESTS

The dangers of low-level radiation from ingested fission products were known to the federal government even before the first atom bomb was built. In *The Making of the Atomic Bomb,* Richard Rhodes relates how Enrico Fermi suggested to Robert Oppenheimer in 1943 that the German food supply could be poisoned with radioactive fission products should building a fission bomb prove impossible. Acting on Fermi's suggestion, Oppenheimer and Edward Teller identified strontium-90 as the isotope that "appears to have the most promise," because it concentrates "dangerously and irretrievably" in human bones. According to Rhodes, Oppenheimer decided "not [to] attempt [such] a plan unless we can poison food sufficient to kill half a million men."[117]

Reflecting on the legacy of the subsequent atmospheric bomb tests, Dr. John Gofman, who headed the biomedical division of the Lawrence Livermore Laboratory and helped develop the atomic bomb, said:

> There is no way I can justify my failure to help sound an alarm over these activities many years sooner than I did. I feel that at least several hundred scientists trained in the biomedical aspect of atomic energy—myself definitely included—are candidates for Nuremberg-type trials for crimes against humanity for our gross negligence and irresponsibility. Now that we know the hazard

of low-dose radiation, the crime is not experimenta-
tion—it's murder.[118]

There is much evidence that government policy, not simple negligence, covered up the adverse health impacts of atmospheric testing. As Bill Curry of *The Washington Post* reported:

> *Officials involved in U.S. bomb tests feared in 1965 that disclosures of a secret study linking leukemia to radioactive fallout from the bombs could jeopardize further testing and result in costly damage claims... That study, as well as a proposal to examine thyroid cancer rates in Utah, touched off a series of top-level meetings within the old Atomic Energy Commission over how to influence or change the two studies. The document also indicates that the Public Health Service, which conducted the studies, joined the AEC in reassuring the public about any possible danger from fallout.*[119]

The Natural Resources Defense Council has recently estimated that, based on a detailed analysis of seismic records, U.S. bomb tests in the period 1945–1962 released the equivalent of 137,000 kilotons of explosive power. The Soviet Union, by setting off several large H-bombs in 1961 and 1962 with 402,000 kilotons of explosive power, accounted for three quarters of the combined total of 585,000 kilotons. Dividing this figure by the estimated power of the Hiroshima bomb yields the fact that the super powers subjected the populations of the world to the fallout equivalent of 40,000 Hiroshima bombs during this 17 year period.[120]

Death rates generally decline over time, driven down by improvements in nutrition, sanitation, and medicine. The trend can be regarded as an inverse indicator of general well being: as mortality declines, life expectancy increases. Despite occasional spikes from wars and epidemics, the long-term decline remains one of the best indices of the advance in health and living conditions. And the continuing decline of mortality indicates that advances in life expectancy will persist in the future.

As shown in Figure 7-1, however, this index changed ominously during the 1950s and 1960s, when both infant and total mortality flattened out.[121] There have been few, if any, epidemiologic investigations of this disturbing phenomena, yet an explanation may be found in the impact of the atmospheric tests.

Rachel Carson was among the first to realize that the sudden emergence of such massive amounts of man-made ionizing radiation could make toxic chemicals even more dangerous. In *Silent Spring,* she wrote:

The most alarming of all man's assaults upon the environment is the contamination of air, earth, rivers, and the sea with dangerous and even lethal materials. . . In

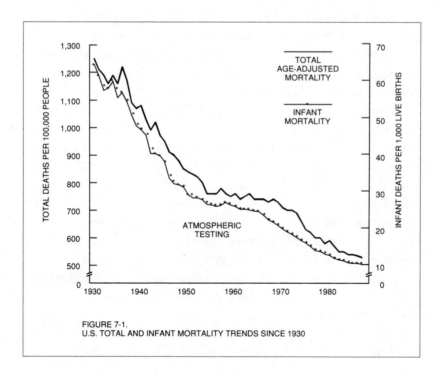

FIGURE 7-1.
U.S. TOTAL AND INFANT MORTALITY TRENDS SINCE 1930

this universal contamination of the environment, chem-
icals are the sinister and little-recognized partners of
radiation in changing the very nature of its life. Stron-
tium-90, released through nuclear explosions into the
air, comes to earth in rain or drifts down as fallout,
lodges in soil, enters into the grass or corn or wheat grown
there, and in time takes up its abode in the bones of a
human being, there to remain until his death.[122]

Though Carson was mindful of the perilous interaction of radioactivity and toxic chemicals during the postwar period, radioactivity rather than toxic chemicals deserves primacy. The cumulative output of organic chemicals in the U.S. rose 42-fold from 1945 to 1965 (from 7.5 million tons to 316 million tons).[123] By contrast, cumulative yield released to the stratosphere rose 13,000-fold (from 45 kilotons to 587 megatons).[124]

Andrei Sakharov predicted in 1958 that 50 megatons of atmospheric bomb tests would cause one half million to one million deaths worldwide.[125] With great prescience, Sakharov anticipated the discoveries made over a decade later by Dr. Abram Petkau of Atomic Energy of Canada, Ltd. concerning the lethal impact of ingested fission products. In 1972, Dr. Petkau demonstrated in the laboratory that low-level radiation generated highly toxic charged oxygen molecules known as "free radicals" that can destroy cell membranes much more efficiently at low dose rates than at high ones.[126] A few years earlier, Dr. T. Stokke and his colleagues observed that very small doses of strontium-90 were much more efficient in damaging bone marrow cells of rats than were high doses.[127] Petkau's discovery could explain surprisingly great levels of immune system damage from protracted exposures to very low levels of fallout radiation, compared with the effect of short intense exposures to X-rays or to the pulse of gamma rays from atomic bomb detonations.

If applied to the fission products associated with the Soviet H-bomb tests of 1961 and 1962, which NRDC estimated were equivalent to 402 megatons of explosive power, Sakharov's estimate would yield four to eight million deaths. At the conclusion of his 1958 paper, Sakharov asked, "what moral and political conclusions can be drawn on the basis of the above figures?" His answer:

> *We are adding to the world's toll of suffering and death . . . hundreds of thousands additional victims, including some in neutral countries and in future generations. The suffering caused by the tests . . . follows immutably from each burst. . . . All the moral implications of this problem lie in the fact that the crime cannot be punished (since it is impossible to prove that any specific human death was caused by radiation) and in the defenselessness of future generations against our acts. The cessation of tests will lead directly to the saving of the lives of hundreds of thousands of people.[128]*

Given this strong admonition, it is easy to understand why the inventor of the Soviet H-bomb fell from grace after the massive Soviet tests of 1961 and 1962. Also in 1958, Linus Pauling speculated:

> *The bomb tests carried out so far [some 150 megatons] will ultimately produce about 1 million seriously defective children and about 1 million embryonic and neonatal deaths, and will cause many millions of people to suffer from minor hereditary defects.[129]*

How accurate were the predictions made by Carson, Sakharov, and Pauling three decades ago?

From 1930 to 1950 total U.S. mortality improved on average by two percent annually (after adjusting for an aging population); however, this improvement fell to only 0.8 percent during the bomb test years. This forboding shift was noted, but not explained, by the Public Health Service in 1964.[130]

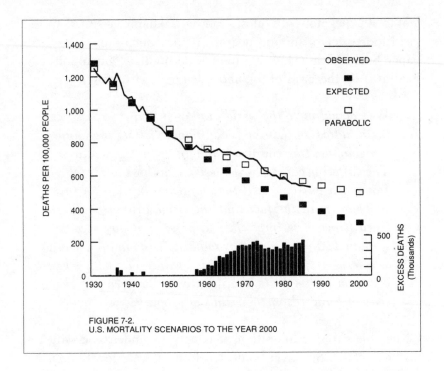

FIGURE 7-2.
U.S. MORTALITY SCENARIOS TO THE YEAR 2000

Figure 7-2 projects two scenarios for future U.S. age-adjusted mortality rates based on the past trend. The first scenario assumes that the age-adjusted mortality rate would continue to decline by two percent each year, as was true prior to 1950. The resulting trend is "asymptotic," meaning the mortality rate keeps approaching zero but never reaches it. This scenario projects "expected" values that fit the observed values quite well during the years up to 1950, but fall well below the observed values in later years. The cumulative difference between these expected rates and those actually observed comes to some nine million excess deaths. The bars at the bottom of Figure 7-2 indicate the annual levels of significant excess deaths, based on this first scenario, a curve that fits observed mortality from 1930 to 1950.

Figure 7-2 also includes a "parabolic" scenario, which was fitted to all of the observed values before and after 1950, and more closely describes the actual trend.[131] Unlike the first scenario, which projects a constant decrease in mortality over time, this scenario projects that an absolute increase in age-adjusted mortality would begin around 2040. (Unlike an asymptote, a parabola's slope eventually reverses, in this case turning upward.) If the post-1950 shift in mortality rates is regarded as natural, then this second projection suggests that the U.S. age-adjusted mortality rate will reach its lowest point in the beginning of the next century, after which it will slowly rise, cancelling out the gains of the past. This scenario suggests that in the future, the prospects for longer life diminish.

But the flattening out of mortality rates in the fifties may not be as natural a development. It could also be the result of a catastrophic environmental error. The former view paints a bleak picture of the future. The latter view suggests a reversal is possible; recognizing the fact of or cause of nine million excess deaths may be the price for future improvements.

From 1915 to 1985, the U.S. rate of infant mortality has improved from about ten percent of babies dying within a year to about one percent. This continued the long secular decline in infant mortality from at least the 18th century when about half of all newborns died within the first year.

But two facts mar this apparent good performance. As with total mortality, infant mortality's four percent average annual decline during 1915–50 flattened out in the period of heavy fallout, resuming only after the test ban was signed. And after 1950, there was a menacing rise in the percentage of low birth-weight babies that continues to this day.

These facts were noted in a Harvard School of Public Health study published in 1968:

> At mid-century there was little reason to expect a major change in the downward trend in [infant] mortality, and additional impressive gains might have been expected. Events have not confirmed this expectation. By 1965 it was possible to look back over a fifteen year period in

which there had been no sizable decrease in the infant
mortality rate. During the 1950s there were years in
which the rate increased—a most unusual occurrence in
half a century of vital statistics reporting in the United
States.[132]

The Harvard study highlighted neonatal mortality (death in the first month of life), which experienced an even more adverse change. It concluded with the hope that increased expenditures on maternal and infant care would correct the problem, but without explaining its cause.

One clue to the problem is the steady increase in the percent of babies born underweight (less than 2,500 grams, or 5.5 pounds), a large contributor to neonatal mortality and miscarriages. The increased rate of low birth weights was especially pronounced among nonwhite babies, whose infant mortality actually rose during this period. From 1950 to 1963, the percentage of very-low weight babies (less than 1,500 grams) increased by about ten percent: the corresponding increase for nonwhite babies was about 50 percent. As stated in an *American Journal of Public Health* article, "an interesting and unexplained observation is that the low birthweight rate and the very low birth-weight rate in nonwhite live births increased steadily from 1950 to the late 1960s, and have been declining slowly since then."[133]

There are numerous socioeconomic causes for infant mortality and low-weight births; however, the timing of the trend suggests the bomb tests may have played an important—and possibly synergistic—role. Figure 7-3 shows how throughout the bomb-test years, rates of nonwhite low birth-weight babies rose dramatically, only to fall again after the cessation of U.S. and U.S.S.R. atmospheric bomb tests. Improvements in neonatal mortality also picked up after the mid-1960s, but this was due primarily to heroic medical efforts to salvage newborn underweight babies.

The surprising vulnerability of young adults to the Chernobyl fallout (see Chapter Two), dramatized by the astonishing increase in AIDS-related deaths during May of 1986, suggests that human immune systems were damaged when atmospheric testing peaked in the 1950s when these young adults were born.

Since 1983, the poorest mortality improvement of any age group was experienced by people 15 to 44 years old, who were infants and children during the bomb test years.[134] (Chinese atmospheric testing did not end until 1980.) These young adults may have sustained immune system damage in those years that is only now beginning to emerge. From 1983 to 1988, while the national age-adjusted mortality rate improved annually, mortality for young adults aged 15 to 44 years old deteriorated each year, the only age group that registered a steady deterioration. There has never been a comparable five-year period in the available historical record in which the mortality rates of young persons increased among both sexes and for whites as well as nonwhites.[135] In fact, a 1984 Department of Energy report states that there have been "slightly higher values [of strontium-90 in human bone] for young adults in New York in the last several years," noting moreover that, "these individuals were children during the period of greatest strontium-90 deposition."[136]

Figure 7-4 traces the change in mortality from disease for eight successive cohorts born since 1920 at two stages of life: age five to nine, when mortality is at its lowest lifetime level, and age 25 to 29, when sexual activity heightens the risk of contracting infectious diseases.

As expected, the lines generally shift downward as mortality rates improved for successive cohorts. However, there were also a significant and unexpected change in the slopes of the lines over time. The lines slope downward for cohorts born prior to 1940, indicating that their mortality rates as young adults were better than as young children. The lines for cohorts born after 1940, on the other hand, slope upward, indicating that their mortality rates were considerably worse as young adults than as children.

The failure of post-1940 young-adult mortality to improve relative to child mortality suggests that some new factor may have

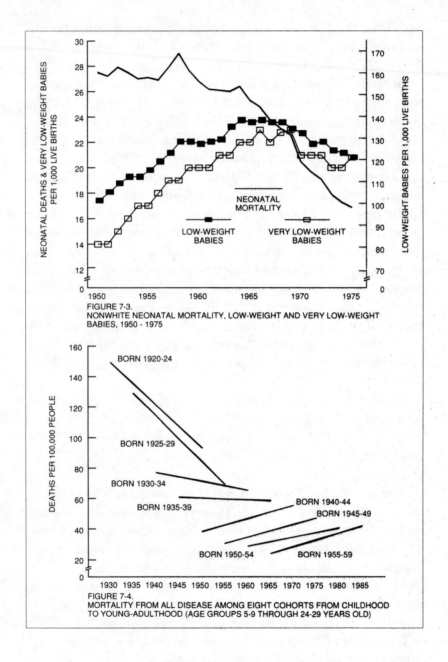

FIGURE 7-3.
NONWHITE NEONATAL MORTALITY, LOW-WEIGHT AND VERY LOW-WEIGHT
BABIES, 1950 - 1975

FIGURE 7-4.
MORTALITY FROM ALL DISEASE AMONG EIGHT COHORTS FROM CHILDHOOD
TO YOUNG-ADULTHOOD (AGE GROUPS 5-9 THROUGH 24-29 YEARS OLD)

damaged immune systems at birth or in early childhood, emerging when the risks of infection peak years later. Unlike the cohorts born prior to 1940, those born afterwards were exposed to fallout from atmospheric bomb tests.

Table 7-1 indicates how both sexes and both official race categories (whites and nonwhites) share this pattern of relatively higher young-adult mortality among cohorts born after 1940, but with some interesting divergences. Adult mortality deteriorated faster among males born after 1940 than among females, possibly reflecting greater sexual activity among males, which plays a role in the transmission of infectious disease. Unlike the case for whites, nonwhite young-adult mortality was no better than nonwhite child mortality prior to 1940, possibly reflecting the damage that poverty, poor nutrition and inadequate prenatal care can do to developing immune systems. Of the four race-sex groups, white males experienced the greatest change in the ratio of young-adult to child disease mortality, followed by white females and nonwhite males, and the least change was among nonwhite females.[137]

Another explanation for the changing ratios of young-adult to child mortality is that improvements in prenatal, infant and child health care have had greater effect on child mortality than any comparable health care improvements have had for young adults. Less health-care improvement for nonwhite children could explain their smaller changes. Greater effectiveness of health-care improvements for male children, who have higher mortality rates than female children, could also explain their larger changes. This reasoning, however, fails to explain why young-adult mortality has actually increased among certain successive cohorts, such as white males.

There are other indicators of possible immune system damage to children born in those years. Figure 7-5 shows that child cancer mortality (age five to nine), which was relatively rare in the U.S. prior to 1945 with rates of less than 20 deaths per million, rose to epidemic proportions during the bomb test years, to a peak of 80 in 1955 in the early days of hydrogen bomb tests. A similar pattern

TABLE 7-1.

CHANGES IN AVERAGE MORTALITY RATES
FROM DISEASE
OF YOUNG ADULTS RELATIVE TO CHILDREN
BORN PRIOR TO AND AFTER 1940

(DEATHS PER 100,000)

	WHITES		NONWHITES		ALL PERSONS
	MALES	FEMALES	MALES	FEMALES	
COHORTS BORN 1920–39					
Mortality Age 5–9	104.4	93.1	143.6	133.4	114.1
Mortality Age 25–29	43.9	56.5	159.2	188.4	71.7
Ratio of Change	0.42	0.61	1.11	1.41	0.63
COHORTS BORN 1940–59					
Mortality Age 5–9	31.2	30.7	42.5	37.7	28.7
Mortality Age 25–29	48.0	48.7	119.2	85.6	48.7
Ratio of Change	1.54	1.59	2.80	2.27	1.70
CHANGE IN RATIOS OF CHANGE FROM COHORTS BORN BEFORE 1940 TO COHORTS BORN AFTER 1940	3.66	2.61	2.53	1.61	2.70

occurred in Japan, where child cancer mortality for males (age five to nine) rose 10-fold from 1945 to 1965.[138]

In 1958, F. Macfarlane Burnet, the distinguished Australian physician and Nobel Laureate, observed that the rapidly growing tissues of children created the best conditions for malignant cells to develop. He also noted that unlike any other disease, the worldwide increase in leukemia was most common among children three to four years old. Dr. Burnet concluded that, "the peak between three and four years of age can hardly have any other interpretation than exposure of the young organism to a mutagenic stimulus around the time of birth."[139]

Figure 7-6 shows a tremendous increase in young-adult septicemia (age 30–34), or blood poisoning—the quintessential immune deficiency disease. Young-adult septicemia was too low to be recorded prior to 1945, but later rose rapidly for persons born from 1950 to 1954, to a peak of seven per million in 1985.

This evidence raises questions about the underlying cause of the wide range of immune deficiency illnesses that young adults suffer today, but which were either rare or totally unknown prior to 1945—diseases such as AIDS, herpes, Chronic Epstein Barr Virus (CEBV, popularly known as "yuppie influenza"), toxic shock syndrome, and deaths from ectopic pregnancies associated with pelvic infections. Young people today are also plagued with *Candida Albicans* infections, which cause birth defects as well as severe discomfort and the appearance of drunkenness. The Japanese first noticed it among atomic bomb survivors and American servicemen after the Second World War, calling it "the drunken illness." Some Japanese scientists believe that the wild proliferation of the *Candida Albicans* organism was caused by mutations from the 1945 atomic blasts.[140] Chapter Ten considers the possible effects of radiation-induced mutations on the organisms that now account for the three fastest growing diseases in the U.S. today—AIDS, CEBV, and Lyme disease.

All of this evidence is in accord with studies done in the early 1970s by Lester Lave[141] and Ernest Sternglass[142] on the adverse

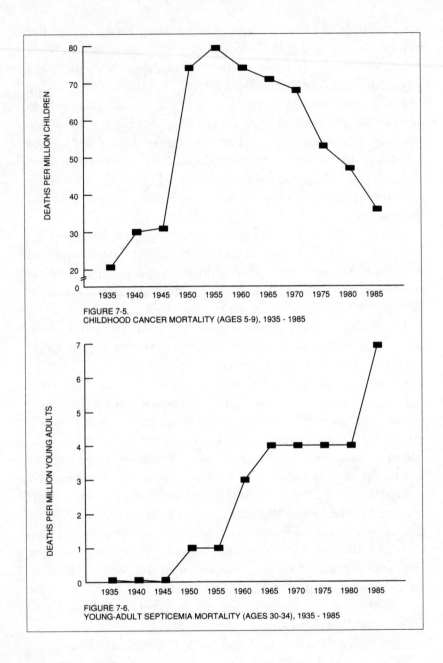

FIGURE 7-5.
CHILDHOOD CANCER MORTALITY (AGES 5-9), 1935 - 1985

FIGURE 7-6.
YOUNG-ADULT SEPTICEMIA MORTALITY (AGES 30-34), 1935 - 1985

impact of fallout in the diet on infant and adult mortality. Further research is urgently needed to evaluate the hypothesis that low-level radiation from fallout is a significant factor in the baby-boom generation's immune-system damage. Such investigations could compare the effects on population cohorts in areas that received the heaviest bomb-test fallout against those in areas with the least exposure. If the hypothesis is correct, postwar immune-system damage would be greater in areas that were hit most heavily by fallout in the rain from the atmospheric bomb tests.

CHAPTER 8

INFANT MORTALITY AND MILK

On March 31, 1987, the Nuclear Regulatory Commission shut down the Peach Bottom nuclear reactors in Lancaster, Pennsylvania and fined the utility over a million dollars, because operators were "sleeping on the job and taking drugs."[143] The industry-backed Institute of Power Operations characterized the reactors as an "embarrassment to the industry."[144] Yet little was said regarding the consequences of the negligent behavior.

One month before the NRC took action, infant mortality in the District of Columbia exceeded the national norm by the greatest margin ever recorded since World War II. High levels of iodine-131 in D.C.'s milk suggest Peach Bottom may have released radioactive emissions under cover of Chernobyl's fallout. Peach Bottom is just 35 miles south of Three Mile Island and just upwind of the largest milk-producing county in this country. Nationwide, nuclear reactors have often been located where milk destined for major cities is produced, because farmland is often relatively inexpensive and underpopulated. Milk contaminated with low-level radiation may be an overlooked contributor to the high urban infant mortality rates. The month after Peach Bottom closed, D.C.'s infant mortality dropped to the national norm for the first time since the plant began operating in the mid-1960s.

After the Chernobyl fallout passed over the U.S., the EPA took very careful measurements of radioactive iodine-131 in milk. Milk is well known to concentrate certain deposited radioactive particulates, and babies and pregnant mothers drink it, exposing infant and fetal thyroids to the deadly contaminants. Iodine-131 increases the risk of thyroid cancer and disease. Doses to infant and fetal thyroids can be hundreds of times more destructive than to adult thyroids, increasing the risks of miscarriages, low birth weights, and infant mortality. Very small doses, moreover, can affect the production of critical growth hormones, disturbing the physical and mental health of children who survive.

Figure 8-1 shows that Washington, D.C., followed by Baltimore, had the highest readings of radioactive iodine-131 in milk of any city on the East Coast in May 1986 when the Chernobyl fallout came—more than twice the levels found in any other Middle-Atlantic city.

It is unlikely that these abnormally-high rates were solely the result of the Chernobyl fallout. The highest Middle-Atlantic read-

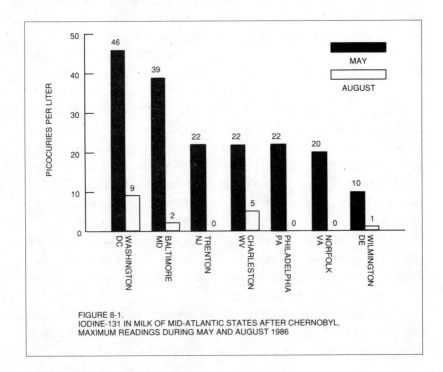

FIGURE 8-1.
IODINE-131 IN MILK OF MID-ATLANTIC STATES AFTER CHERNOBYL,
MAXIMUM READINGS DURING MAY AND AUGUST 1986

ings of iodine-131 in the rain, which brought down the Chernobyl radiation, were found in Virginia.[145] But Virginia's milk readings were less than half those in D.C. Moreover, D.C.'s milk continued to have high radioactive iodine levels in August, three months after the Chernobyl fallout passed by. Because of its short half-life of only eight days, the Chernobyl iodine-131 would not have been detectable by August. Only a reactor release could have caused such a high reading in August of 1986.

Why was D.C.'s milk so contaminated? Iodine-131 does not occur naturally; its only possible source is a nuclear reactor. The data suggest that the Chernobyl cloud may have obscured radioactive emissions from Peach Bottom. Three Mile Island, located just 35 miles north of Peach Bottom, may have also had emissions; however, TMI was under careful scrutiny during clean-up, while Peach Bottom was reportedly operating in a careless manner. Other nuclear plants near D.C., such as the Calvert Cliff plant on Maryland's Chesapeake Bay shore, are also unlikely to have caused the high milk contamination, since little milk is produced in the counties surrounding them.

A large release from Peach Bottom also explains why Philadelphia and New York City had such high iodine-131 readings in milk in May 1986.[146] Both cities receive milk from the counties within a fifty-mile radius of Peach Bottom. Federal Milk Marketing Administration records indicate significant amounts of milk and dairy products are shipped from these counties to the New York metropolitan area, as well as to the Middle Atlantic Milk Marketing Area, which includes Philadelphia, Baltimore and D.C. Smaller cities with surrounding dairies further than 50 miles from Peach Bottom, on the other hand, had lower concentrations of iodine-131 in their milk. This pattern is consistent with a localized release from Peach Bottom, but not with exposure to Chernobyl fallout.

The Peach Bottom plant has long been plagued with serious problems. Its first reactor began operating in 1966, but leaked so badly that it had to be decommissioned a few years later. Units Two and Three started in the early 1970s, and they too leaked large amounts of radioactive fission products within the next few years. The NRC issued a series of warnings that release limits were violated and required extensive modifications of the radioactive waste treatment and filtration systems. After a major release in 1976, however, the NRC simply provided a three-fold increase in the amount of radiation Peach Bottom was allowed to release![147]

Peach Bottom Units Two and Three are boiling-water reactors built by the General Electric Company. This type of reactor is more prone to release radioactive gases under normal operating conditions than are the pressurized-water reactors originally designed by Westinghouse for use in nuclear submarines. Boiling-water reactors use a "single-loop" design; the steam produced by the uranium fuel goes directly to the electric generator turbines. Pressurized-water reactors, on the other hand, have two loops, one for hot water circulating through the hot fuel elements, and another for steam going to the turbines. The second loop helps prevent radioactivity from reaching the turbines, where it can leak through the bearings and escape into the environment. Only pressurized-water reactors are used in submarines, because any radioactive gas leaks would allow the submarine to be detected and tracked. Since enemy detection is not a problem for commercial reactors, the AEC encouraged General Electric to build the cheaper single-loop reactors.

The Peach Bottom reactors continued to perform poorly during the 1980s. In 1983, the plant spilled 2,500 gallons of radioactive water, was found to have cracked cooling pipes, and was fined $140,000 by the NRC for violations made in 1982 and another $40,000 for valve violations. In 1984, the NRC fined the operator, Philadelphia Electric Company (PECO), another $30,000 for violations. In 1985, the plant was shut down because of mechanical malfunctions, and water levels in Unit Two dropped dangerously. Both the NRC and OSHA fined PECO for safety violations, which allegedly exposed workers to radiation and led to the death of an employee. In 1986, control rods were inappropriately withdrawn from the reactor core, the emergency power substation had an explosion and fire, and the NRC condemned PECO for a "pattern of inattentiveness and sloppiness," calling Peach Bottom "one of the worst plants in the nation," reporting 17 violations, and accusing management of illegally dismissing a whistle-blowing health physicist.[148]

The NRC eventually shut the plant down in March 1987. Along with a $1.25 million civil penalty, the NRC took the unprecedented action of fining individual operators. Violations had led to numerous emergency shut-downs, or "scrams," which can produce thermal shocks that release iodine-131, strontium-90, and other fission products from cracks in the fuel rods. In addition to finding operators asleep at the control desk, numerous deficiencies in operator training and discipline were also found. The plant remained closed for some time, pending major management changes, operator retraining, mechanical modifications and repairs mandated by the NRC, but was expected to resume operations as of this writing.

Prior to the start-up of Peach Bottom's first reactor, nonwhite infant mortality rates for Baltimore and D.C. were better than those for the rest of the country. Figure 8-2 shows that throughout the 1950s and 1960s, D.C.'s nonwhite infant mortality was lower than that of the rest of the country, which is shown as the solid baseline equal to one. Starting in the early 1970s, however, D.C. skyrocketed above the rest of the country. Baltimore experienced a similar but less dramatic rise. Maryland and Virginia remained relatively stable, at around the U.S. rate, though they experienced a peak in 1976, when Peach Bottom released a massive 212,000 curies of radioactivity. Only nonwhites are considered here, because there are too few white infant deaths in D.C. for statistically significant determinations.

Figure 8-3 focuses on the large increases in nonwhite infant mortality relative to the U.S. that occurred in D.C. after 1973, when Peach Bottom Units Two and Three and Three Mile Island released large amounts of radioactive gases. Peach Bottom accounted for 80 percent of the half-million curies released to the air from 1973 through 1983, excluding 1979, when the Three Mile Island

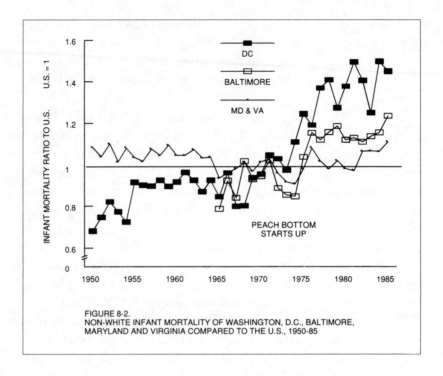

FIGURE 8-2.
NON-WHITE INFANT MORTALITY OF WASHINGTON, D.C., BALTIMORE,
MARYLAND AND VIRGINIA COMPARED TO THE U.S., 1950-85

accident released ten million curies.[149] These releases were well correlated with the peaks in D.C.'s infant mortality, accounting for almost half of the variation observed in the upper graph.

Figure 8-4 shows that there was a statistically significant peak in D.C.'s infant mortality in the months following the high readings of iodine-131 in the summertime milk of 1986. Apparently, many newborn babies were unable to survive the environmental insult.

An even more disturbing event occurred around April of 1987, nine months after the high milk readings were recorded. Babies born then had been first-trimester fetuses when the milk readings were highest. In their first trimester, these babies would have been the most susceptible to contaminated milk consumed by their mothers. In April of 1987, D.C.'s relative rate of infant mortality was the highest recorded since World War II, fully three-and-a-

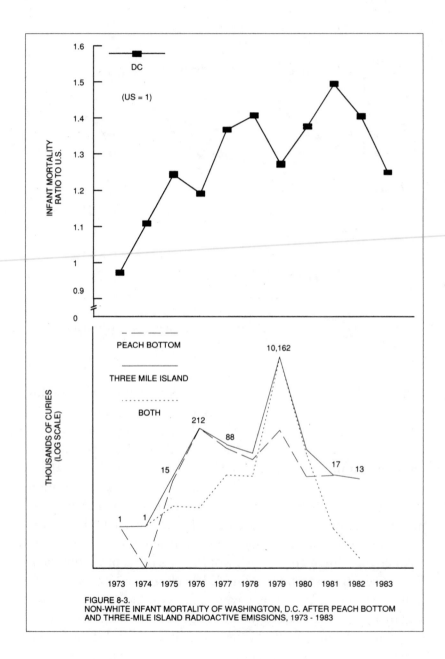

FIGURE 8-3.
NON-WHITE INFANT MORTALITY OF WASHINGTON, D.C. AFTER PEACH BOTTOM
AND THREE-MILE ISLAND RADIOACTIVE EMISSIONS, 1973 - 1983

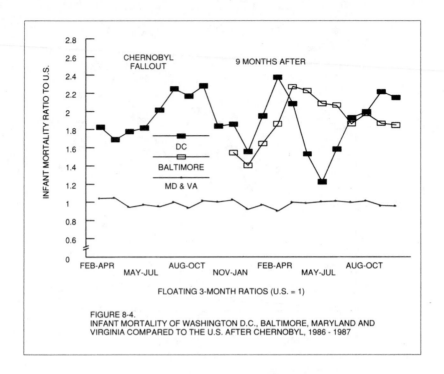

FIGURE 8-4.
INFANT MORTALITY OF WASHINGTON D.C., BALTIMORE, MARYLAND AND
VIRGINIA COMPARED TO THE U.S. AFTER CHERNOBYL, 1986 - 1987

half times higher than the rest of the nation. Baltimore's rate was two-and-a-half times that of the U.S. the following month.[150]

An improvement would be expected after exposure ceases. Sure enough, in May, 1987, the month after Peach Bottom was shut down, infant mortality in D.C. returned to about the same rate as in the rest of the U.S. for the first time since the reactor station began operating in the mid-1960s.[151] Subsequent increases in D.C.'s infant mortality may have been due to the fact that Three Mile Island resumed operations soon after Peach Bottom closed down.

Since there are no dairy cows within the District of Columbia or Baltimore city limits, high iodine-131 readings originate where the milk is produced. Unfortunately, it is rather difficult to trace milk sold in a given area back to the county—let alone the dairy farm— from where it came. With the development of better highways and refrigerated transporting equipment, the distribution of fresh milk is no longer a localized industry. Today, "bulk milk can be shipped more than 1,500 miles to an area short of milk."[152] The federal government began to regulate milk distribution during the 1930s in order to guarantee small dairy farmers a fair price. State and local governments were left with the responsibility of ensuring sanitary and quality standards, so if milk is suspected of causing a localized health problem, it is up to local authorities to determine its source.

For example, the Food Protection Branch (FPB) of D.C.'s Department of Consumer and Regulatory Affairs is responsible for regulating the District's milk supply. The FPB, however, does not keep detailed records of milk sources, determining its origins only when necessary.[153] Earlier records indicate D.C. milk has come from fourteen states, with some as far as California, Texas, and Wisconsin.[154] According to the FPB, eleven Pennsylvania dairies have shipped milk to D.C. in the past, four of which were located within 50 miles of Peach Bottom; however, none would admit shipping to the District when asked.[155] Maryland's Department of Health and Mental Hygiene does keep track of Baltimore's milk sources; however, the information has still not been provided more than a year after the request was originally made.

Short of identifying exact sources for particular cities, statistics on where milk is produced and the general marketing areas to which it is shipped are available from the Federal Milk Marketing Program, which regulates about 80 percent of all grade-A milk produced in the country.[156] These data indicate that it would be difficult to find a large city that does not get some milk from dairies near a nuclear reactor. Reactors have generally been located in rural areas to keep them away from people. But in those same rural areas are farms that produce large quantities of milk. Apparently in selecting specific sites it was not recognized that milk produced by farms near the reactors would be shipped long distances to the urban markets.

New York City, for example, is dependent on milk shipments from the counties of Jefferson, Oswego, Saint Lawrence, and Lewis, which are all downwind of the Nine Mile Point and James A.

Fitzpatrick reactors. Dairy plants in Minnesota and Wisconsin account for almost half of the milk produced in the country and supply most cities of the Midwest. Minnesota's milk-exporting counties of Benton, Isante, Morrison and Sherbourne are just downwind of the Monticello reactor near Saint Cloud. The dairy counties of Dakota, Goodhue, Hennepin, Scott and Washington are downwind of the Prairie Island reactor (26 miles southeast of Minneapolis). Wisconsin's milk-exporting counties of Pierce and Saint Croix are also near Prairie Island. Buffalo, Jackson, La Crosse, Monroe, and Trempealeau produce milk near the La Crosse reactor, and the dairy counties of Brown, Door and Kewaunee are downwind of the Kewaunee reactor (27 miles away from Green Bay).

Four New Jersey counties near reactors ship almost all of their milk to the Middle-Atlantic marketing area, where the cities of Philadelphia, Baltimore and D.C. are located. These milk-producing counties include Burlington, near the Oyster Creek reactor, and Cumberland, Salem, and Gloucester, which are near the Salem reactor. Even in Virginia, which ships very little of its milk out of state, there are two small milk-producing counties near nuclear reactors that ship all of their milk to the Middle-Atlantic market: Louisa, near the North Anna reactor, and Surrey, near the Surrey reactor.

Surrounding Peach Bottom itself is one of the most productive dairy regions in the country. The 15 counties within 50 to 60 miles of the Peach Bottom reactors account for about four percent of all the milk produced in the United States—some four billion pounds per year. Peach Bottom is located at the junction of Lancaster and York counties, on the border of Pennsylvania and Maryland, close to the Susquehanna River which flows past Baltimore into the Chesapeake Bay (see figure 8-5). Lancaster county is the largest milk producing county in the country, producing approximately 1.5 billion pounds annually.

Almost half of all the milk sold in the Middle-Atlantic area comes from within 50 miles of Peach Bottom. Lancaster County

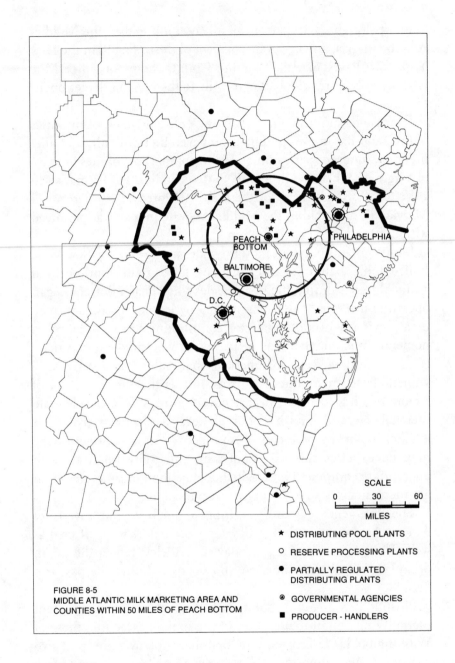

FIGURE 8-5
MIDDLE ATLANTIC MILK MARKETING AREA AND
COUNTIES WITHIN 50 MILES OF PEACH BOTTOM

SCALE

0   30   60

MILES

★ DISTRIBUTING POOL PLANTS

○ RESERVE PROCESSING PLANTS

● PARTIALLY REGULATED
    DISTRIBUTING PLANTS

⊛ GOVERNMENTAL AGENCIES

■ PRODUCER - HANDLERS

accounts for about 16 percent of all the milk sold in the Middle-Atlantic market, and about five percent of the milk sold in the New York-New Jersey market. Dauphin County, home of Three Mile Island, sends more than half of its milk to the New York metropolitan area.

Lancaster's surplus could satisfy the entire demand of the cities of D.C. and Baltimore combined, and even then with almost half a million pounds to spare annually.[157] In comparison, dairies from all of Virginia sold far too little milk to the Middle-Atlantic market in 1987 to satisfy the needs of these cities. Moreover, 40 percent of Maryland's milk sold in the Middle-Atlantic market was produced within 50 miles of Peach Bottom.

Improvements in infant mortality rates in the cities of New York, Philadelphia, Baltimore and D.C. have fallen significantly behind national trends during the two decades since Peach Bottom began operating in 1966. Milk contaminated by low-level radiation may be a previously-overlooked factor contributing to this urban phenomenon, and one that may be particularly detrimental to African Americans. It is well documented that babies are particularly vulnerable to environmental toxins, just as they are to a host of factors relating to urban decay: poverty, drugs, lack of adequate prenatal care, and so forth. Yet urban decay as a comparative factor is offset in part by problems of rural decay. An underweight baby may have better survival chances in inner-city hospitals with appropriate equipment and skilled doctors than in rural areas that are increasingly losing hospitals and physicians.

The divergence cannot be attributed solely to the poverty of African-Americans in these cities, for white babies were affected as well. The excess infant deaths may nevertheless reflect the urban decay associated with inner-city fiscal crises during the past two decades.

Pittsburgh is a case in point, with the highest nonwhite infant mortality rate of all major cities in the nation, exceeding those of Washington D.C., Detroit, and Baltimore for over the past two decades.[158] According to the *Pittsburgh Post Gazette*, "more black women receive proper prenatal care in Pittsburgh than in many cities with lower infant mortality rates." The article goes on to state that:

> *Pittsburgh's high [infant mortality] rate [is] all the more confounding, for Pittsburgh seems healthy according to other indicators, such as drug abuse and crime. It is a pattern occurring in a city famous for saving children's lives with liver transplants and for its advances in multi-organ transplants. . . . It's a city with the lowest number of AIDS cases of any major metropolitan area in the United States. And it's a city that hasn't yet suffered the scourge of crack that other large towns have. And it's a city with a murder rate so low that when Miami police were told that there were 24 homicides in 1988, they asked which month. . . . The conventional wisdom that more prenatal care is the solution for high black infant mortality doesn't seem to hold true for Pittsburgh.[159]*

The *Post Gazette* did not consider the possible connection with consumption of milk from counties such as Beaver, Washington, Lawrence, and Butler, all of which are exposed to emissions from the Beaver Valley reactor in Shippingport, Pennsylvania, as well as from the Lancaster-York area.

Nonwhites have been hardest hit by the poor improvement in infant mortality trends in Pittsburgh and other large cities during the past two decades. If, as the experience of Pittsburgh suggests, poverty and the lack of prenatal care may not be the sole causes, then other causes must be sought. One area which deserves attention is dietary differences between whites and nonwhites. For example, cheese is widely handed out to poor families from government surplus storage. Since milk products such as cheese are also contaminated by the long-lived radioactivity of strontium-90 and cesium-137, many families dependent on government aid may be disproportionately affected.

Aside from milk, whites generally eat more foods like fish and meats which have a higher ratio of calcium per unit of strontium-90 than less expensive root vegetables commonly eaten by African Americans. Consequently, African-American women may build up higher levels of long-lived strontium-90 in their bones, resulting in more vulnerable immune systems.

One newly-discovered consequence is that such a damaged immune system might reject the developing fetus as a foreign body.[160] In such cases infant mortality increases occur within two or three years of exposure to the fission products. The health effects of ingesting milk contaminated with short-lived iodine-131, on the other hand, can occur within a matter of weeks or months after exposure.

These are complex questions which have yet to get the attention they deserve. Unfortunately, in 1974 the EPA ceased publication of annual measurements of strontium-90 in human bone, which would help ascertain whether urban nonwhites do in fact have higher concentrations than whites.

The argument that improvement in prenatal care is all that is needed to offset the deterioration of urban infant mortality rates ignores the fact, mentioned in Chapter Seven, that the rate of low birth-weight babies began to increase in the fifties. Every radiation release examined in this book was accompanied by increases in infant mortality. From the late 1960s to the early 1980s, moreover, improvements in infant mortality rates of the 30 states with nuclear reactors or located just downwind of them fell significantly behind the remaining states.[161]

When examined nationally, there is a disturbing and statistically significant correlation between, on the one hand, regional risks of exposure to milk contaminated with low-level radiation from commercial reactors, and, on the other hand, increases in infant mortality. Figure 8-6 shows how strikingly similar this correlation is to the supralinear dose-response relationship suggested by the Chernobyl data (shown in Figures 2-4 and 2-5). For example, the milk-radiation exposure risk for eight Northern-Midwestern states (Illi-

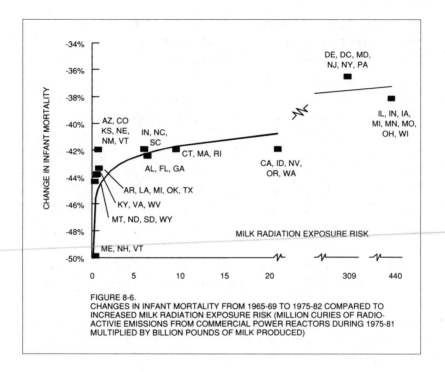

FIGURE 8-6.
CHANGES IN INFANT MORTALITY FROM 1965-69 TO 1975-82 COMPARED TO
INCREASED MILK RADIATION EXPOSURE RISK (MILLION CURIES OF RADIO-
ACTIVIE EMISSIONS FROM COMMERCIAL POWER REACTORS DURING 1975-81
MULTIPLIED BY BILLION POUNDS OF MILK PRODUCED)

nois, Indiana, Iowa, Michigan, Minnesota, Missouri, Ohio, Wisconsin) is estimated to be more than a thousand times greater than that for three Northern New England states (Maine, New Hampshire, Vermont), and infant mortality in these Northern-Midwestern states rose ten percentage points faster, closely fitting the supralinear curve. The probability is less than one in a hundred that this national correlation is due to chance.[162]

Figure 8-6 calculates "exposure risk" by multiplying the total volume of milk produced in each region by the amount of low-level radiation emitted from commercial reactors there during 1975 through 1981. The changes in infant mortality are calculated for the period 1965–69, when few commercial reactors were in operation, to the period 1975–82, when more than two-thirds of all radioactive emissions from commercial reactors occurred.

City residents are victims of limited information on the dangers of low-level radiation in milk. Information on monthly emissions from Peach Bottom, Three Mile Island and other reactors are unavailable, as are adequate data on milk sources. Government officials usually downplay the potential risks of such hazards to avoid controversy. For example, when preliminary findings from this chapter were released at an NRC hearing to reopen Peach Bottom in February 1989, a Maryland Department of Health and Mental Hygiene spokesperson immediately contradicted them, claiming none of Washington's milk came from within 50 miles of Peach Bottom. Yet it turned out that the person relied on for this claim could not provide documented information regarding the sources of D.C.'s milk.

Ignorance has been fostered even more by the permissive limits on radioactivity. In 1964, with no public hearing, the Johnson Administration relaxed an earlier AEC limit on permitted emergency releases of iodine-131, from 500 to 15,000 picocuries. No convincing reasons for this thirty-fold increase were offered; however the use of atomic bombs was being seriously considered to resolve the Vietnam conflict at the time.

There is one way to dispel this ignorance, and at the same time to determine the relative importance of contaminated milk as a factor contributing to the lagging improvements in urban infant mortality. With the restart of Three Mile Island and Peach Bottom reactors, environmentalists in New York City, Philadelphia, Baltimore, and Washington might call for a boycott of fresh milk for one year, substituting it with powdered milk, which normally contains no iodine-131. If infant mortality drops significantly in the ensuing months, as it did in Baltimore and Washington after the shutdown of the Peach Bottom reactors, then the dangers posed by contaminated milk should become clear to all.[163]

CHAPTER 9

CANCER IN CONNECTICUT

By the middle of the 1960s, the nuclear industry expected to build one thousand more nuclear power plants by the turn of the century. The licensing of these plants was proceeding unimpeded, accompanied by public assurances that their electricity would be "too cheap to meter." Among these powerful new reactors were two at Three Mile Island, two at Peach Bottom, and two at Millstone, near New London, Connecticut. These reactors have been among the most troublesome in the nation, with repeated large releases of low-level radiation.

Millstone ranks second to Three Mile Island in emissions of fission products from commercial reactors, having released nearly seven million curies by 1983.[164] About three million curies were emitted in 1975 alone, including ten curies of iodine-131, almost as much as the TMI accident reportedly released. Since 1975, many observers have noted that Connecticut's cancer mortality outpaced the nation's, particularly in areas closest to Millstone and the Haddam Neck reactor, about 25 miles northwest.

Connecticut, home of General Electric, builder of boiling-water reactors, is more reliant on nuclear power than any other state in the country. The Connecticut Department of Health Services has denied that there is a linkage of radiation and cancer, and has curtailed public access to information necessary for conducting an investigation.

As in the case of the TMI accident in 1979 and the Pilgrim accident in 1982, Dr. Ernest Sternglass was the first to discern the association between cancer increases and radiation released from the malfunctioning Millstone reactor. Sternglass found that increases in cancer mortality rates from 1970 (before Millstone began operations) to 1975 varied directly with distance from the reactor.

From 1970 to 1975, cancer mortality increased by 58 percent in Waterford Township, where Millstone is located, and by 44 percent in New London, five miles east of the reactor. Cancer mortality in Connecticut as a whole increased by twelve percent, which was twice as fast as the corresponding six percent U.S. rise. The increases in all New England states downwind of the reactor displayed the same correlation with distance: Rhode Island and Massachusetts rose eight and seven percent respectively, while further away, New Hampshire rose by only one percent and cancer mortality in Maine actually fell by six percent.

Similar strong correlations were also observed between radioactivity in milk during 1976 and distance from the reactor. The peak readings for both radioactive strontium and cesium in all of New England were measured at milk farms closest to the Millstone reactors. No Chinese bomb tests could be blamed for the very high readings near Millstone; although the utility and the NRC claimed this was the case in their reply to Congressional inquiries.

Sternglass delivered the results of his research to then-Congressman Christopher Dodd in 1977, before the huge Millstone release of 1975 was made public.[165] Sternglass had been concerned about cancer increases in Connecticut since Millstone and Haddam Neck reactors first began operating in 1967 and 1971 respectively. Dodd asked the NRC representatives whether they considered the amounts of strontium-90 found in the milk around the Connecticut plants to be unrelated to plant operations. In response, Commissioner Joseph M. Hendrie said the NRC "has not been able to establish any potential pathway of environmental transport from the Connecticut plants to the milk which would account for the levels observed."[166]

Since 1978, environmental and citizen groups have continued to raise questions about the health effects of Connecticut's reliance on nuclear power. In 1987, the Connecticut Citizen's Action Group (CCAG) invited Public Data Access, Inc. (PDA) to update the Sternglass findings, in the light of the new information that the Millstone releases of 1975 represented the second largest in the history of civilian reactors. The results were presented at a March 1987 news conference in Hartford, which demonstrated that from the late 1960s to the early 1980s, cancer mortality rates in Connecticut increased significantly faster than in the rest of the U.S. The increase was significantly greater in the four counties closest to the reactor, including Middlesex and New London in Connecticut, and Kent and Washington in Rhode Island.[167]

But data were not available for the townships of Waterford and New London, which Sternglass had noted had the largest cancer mortality gains from 1970 to 1975. Township cancer mortality rates had been routinely published by the Connecticut Department of Health since the 1930s, but were suddenly omitted from annual reports published after 1977, without explanation. The data for 1977 further revealed the importance of township cancer mortality rates for analyzing the effects of the Millstone emissions; since 1970, cancer deaths increased by 62 percent in Waterford and by 45 percent in New London, compared to an 18 percent rise in Connecticut as a whole.

In response to PDA's findings, Richard Gruber, head of the Connecticut Department of Health Services, said that age-adjusted cancer mortality had only increased by 3.1 percent from 1970 to 1980.[168] By suggesting that the increases in the crude mortality rates merely reflected faster aging of Connecticut's population, Gruber avoided explaining the divergent trends found in areas closest to the reactor.

State health departments generally respond to analyses of their published crude mortality rates by referring to unpublished age-adjusted data, which are unavailable without laborious and expensive computerized adjustment procedures. For this reason, PDA has spent much time and money developing state and county age-adjusted mortality rates; however, these data were not available at the time of the Millstone investigation in 1987.

Paul Gionfriddo of the Connecticut Legislature's Public Health Committee was sufficiently disturbed by PDA's findings to set aside $25,000 for a study to be conducted jointly by PDA and the Department of Health Services. The proposed study would use the Department's age-adjusted township mortality data, with emphasis on areas closest to the reactors. At a May 1987 meeting, however, Dr. Gruber said there was no need for PDA's participation, and declined to use PDA's database on toxic pollution and other possible confounding factors in the study.

The health department's report was delivered to the Public Health Committee on December 31, 1987, with the following summary and conclusion:

> The study ... does not address the broader research question of cause and effect relationships between low-level radiation and cancer risk. The study did examine cancer rates for five specific types of cancer known to be sensitive to ionizing radiation, and focused primarily on incidence rates adjusted for demographic factors. Leukemia was of special interest because the time from exposure to diagnosis is shorter by several years than it is for solid tumor development. Proximity to the Haddam Neck and Millstone facilities, as well as rainfall patterns were also examined.
>
> In light of both the lack of statistical significance in the results and their lack of indication of a meaningful pattern of rates, we conclude that there is not sufficient evidence to warrant a more detailed, labor intensive investigation.[169]

There are several questions raised by this study which are extremely puzzling. Instead of using the Health Department's age-adjusted cancer mortality figures for townships near the reactors, the study used cancer incidence and mortality data from the Connecticut Tumor Registry, an independent body of data that traces the history of cancer patients back as far as 1935. But the study did not offer any mortality estimates that could be used to compare the tumor registry's coverage with the Department's own cancer mortality data.

The completeness of the tumor registry has been questioned, however. The *Hartford Courant*, for example, reported that a Yale-New Haven hospital study of lung cancer autopsies over a ten-year period concluded that "the lung cancer rate may be four times higher among Connecticut men and 15 times higher among the state's women than is reported in the state's Tumor Registry."[170]

Not only is it difficult to perform an independent check of the data used in the study, but its methodology was also severely flawed. The study excluded any city that happened to fall on boundaries drawn arbitrarily at ten and 20 miles around Millstone and Haddam Neck. By so doing, the Health Department failed to consider such "border towns" as Hartford and Manchester! Only a handful of rural townships close to the reactors were studied instead, accounting for only eight percent of the state's population, thus ensuring a sample too small to provide statistically meaningful results.

The University of Connecticut has published a cancer atlas that provides evidence contradicting the state's inconclusive results.[171] The atlas uses the tumor registry's age-adjusted cancer data for the years 1958 to 1982, grouped into 14 areas. According to the atlas, the cancer incidence in the areas surrounding Middletown and Groton, which are the two closest to Millstone, was increasing significantly faster than in the state as a whole.[172]

Dr. Holger Hansen, the Director of the University of Connecticut cancer atlas, has also reported alarming increases in cases of Down's Syndrome and other birth defects in recent years, as compared with the period 1970–72, suggesting that "maybe something very unusual is happening in Connecticut."[173] This report takes on added significance in light of the discovery, detailed in Chapters Four and Five, of highly significant increases in birth defects after the radiation releases in the aftermaths of the Savannah River accidents of 1970 and the Three Mile Island accident of 1979.

The Connecticut media has devoted substantial coverage to the widespread concerns about low-level radiation among state residents. A random telephone survey of 243 homes in Windham conducted by the Connecticut Citizen's Action Group revealed that thirty-nine percent of the respondents agreed that "nuclear power is so hazardous that the power plants in Connecticut should be shut down."[174]

Residents are also worried about the rapid increase in the incidence of Lyme disease, named after the town of Old Lyme, just ten miles from Millstone. A recent article on Lyme disease notes that:

> *Lyme disease was first reported in November 1975, when the Connecticut State Health Department received telephone calls from two mothers whose children had just been diagnosed as having juvenile rheumatoid arthritis. The condition is a devastating one that can lead to lifelong suffering and physical debilitation, and it is not surprising that these mothers were concerned. What alarmed Health Department officials, however, was the news that these cases were not isolated ones. According to the women who telephoned, a number of adults and children in the town of Lyme had recently been diagnosed as having rheumatoid arthritis. Health officials concluded that this was possibly the beginning of an epidemic.[175]*

Lyme disease has since spread rapidly. In 1975, 59 cases were recorded in Connecticut; in 1985, the number increased to 863, mainly in the two counties of Middlesex and New London. Just as increases in cancer may be linked to the huge radiation release from Millstone in 1975, so too may be the tick-borne Lyme disease epidemic. The Lyme disease is carried by a spirochete that had not been harmful to humans prior to 1975. It is well known that radiation can cause mutations in bacteria. The 1975 Millstone radiation release may have caused just such a mutation in the tick-borne spirochete, and may have unleashed yet another health problem on the public.

Such speculation will undoubtedly be dismissed by some as ungrounded and irresponsible, just as Sakharov's 1958 warnings of epidemics from fallout-induced mutations had been ignored in the Soviet Union. But the cause of this alarming outbreak can only come to light if such hypotheses are raised and tested. Important clues to the validity of such hypotheses would come from the publication of Connecticut Department of Health data on cancer and Lyme disease morbidity and mortality by township. There remains the need for an objective study of the change in excess death from cancer and other diseases in Connecticut since 1975.

CHAPTER 10

RADIATION AND AIDS

Among the most rapidly growing diseases today are AIDS, Chronic Epstein Barr Virus, Lyme disease, *Candida Albicans*, herpes, septicemia and several other immune-deficiency ailments mentioned elsewhere in this book. The rise of these maladies in recent years, particularly among young people born in the nuclear age, may be related to the huge amounts of low-level radiation released since 1945. The research necessary to test this hypothesis has barely begun. This chapter reviews some of the research and offers suggestions on how to continue.

Chapter Seven established the signal fact that young-adult mortality from disease among generations born in the Nuclear Age has deteriorated considerably relative to childhood rates. This is in sharp contrast to generations born in the 1920s, who were not exposed to low-level radiation as babies and children, and whose young-adult mortality actually improved compared to childhood rates. This finding supports the hypothesis that fission products entering into the food chain may have damaged the immune systems of a significant proportion of the baby-boom generation.

Dramatic increases in young-adult mortality from septicemia, that is, blood poisoning associated with immune deficiency, is further support for this hypothesis, as noted in Chapter Seven (see figure 7-6). The federal government now lists septicemia among the top dozen causes of death, with a 1987 mortality rate of 8.1 deaths per million people. Septicemia mortality increased by an astonishing 15 percent annually since the 1950s, when the rate was only 0.3 deaths per million people. If the ten percent yearly gain observed since 1970 persists, septicemia will kill 76,000 people annually by the year 2000.

The only comparably rapid rise has been in the AIDS mortality of recent years. The National Center for Health Statistics (NCHS) estimates there were 13,120 AIDS deaths in 1987. These deaths are included in the category "deaths from all other infectious and parasitic diseases," which has been increasing by about twenty-two percent annually since 1979. If this yearly increase stabilizes at ten percent, as many as 46,000 people will die annually from AIDS-related diseases by the year 2000.

These are truly frightening figures. It is surprising that so little attention has been paid to the emergence of septicemia as an immune deficiency disease that outranks AIDS in relative importance. But the danger posed by immune-system deficiencies goes beyond just AIDS and septicemia, for they diminish our ability to cope with all illnesses.

Heart disease, for example, accounts for roughly half of all deaths each year. Mortality from heart disease has been declining by about 1.3 percent per year since 1970, reaching a 1987 rate in of 397 deaths per 100,000 people. Yet its decline was interrupted twice. From 1971 to 1972, heart disease mortality increased from 493.3 to 497.1 deaths per 100,000 people, and again from 1979 to 1980, the rate rose from 426.7 to 436.4 deaths per 100,000 people. Chapter Five notes that during these years the age-adjusted mortality rate in the whole nation rose significantly. The years 1971 to 1972 were in the wake of the Savannah River Plant accident, and 1979 to 1980 right after the Three Mile Island accident. It is important for future research to determine whether the unusual increases in heart disease mortality during the two periods were centered in areas most affected by the accidental radiation releases.

Patients also depend greatly on their immune systems for recovery from pneumonia. U.S. mortality from pneumonia rose significantly after Chernobyl fallout arrived in the U.S.; during the period

1971 to 1972 after the Savannah River accidents, and during the period 1979 to 1980, after TMI. Even more disturbing, pneumonia mortality rose 38 percent from 1982 to 1987. This was the first such long period since the 1950's in which the U.S. crude mortality rate failed to improve. It was also the first time ever that mortality rates for the age groups from 15 to 44 years failed to improve, as noted in Chapter Seven. These may all be examples of immune system damage traceable to atmospheric fallout, and exacerbated by later accidental releases of fission products from nuclear reactors and reprocessing facilities.

African-American male mortality rates from immune deficiency diseases has been consistently higher than those for other people. According to National Center for Health Statistics, for example, African-American male age-adjusted mortality from septicemia was 11.9 deaths per million people in 1985, nearly three times the rate for all people. African-American males' average lifetime is also deteriorating, declining from 65.6 years in 1984 to 65.4 years in 1987.[176] And nonwhites were found to suffer greater mortality peaks after the Savannah River Plant accidents than whites.

Since Nobel Prize-winner Herman Müller began experimenting with fruit flies in the 1920s, radiation has been known to accelerate the mutation of organisms. Over the past half century, radiation may well have created many new organisms that can take advantage of weakened immune systems.

For example, there has been an enormous increase since the mid-fifties in the number of insect and mite species that have become resistant to pesticides. In 1938 there were only seven such organisms known, and in 1955 only 25. But by 1984 the number climbed to 447, a growth rate ten times faster than the 1938 to 1955 trend.[177] As a result, despite the greater complexity and cost of modern chemical pesticides and far heavier applications,

> pests have evolved mechanisms of detoxifying and resist-
> ing the action of chemicals designed to kill them. . . .
> Resistance in weeds was virtually nonexistent before

*1970. But since then, with the growth of herbicide use, at
least 48 weed species have gained resistance to chemicals.
. . . Farmers and pesticide producers have locked them-
selves into a race with the rapid evolution of crop pests.*[178]

The author of this study offered no explanation for the extraordinary "evolutionary" capacity of these organisms after the mid-
fifties, and did not consider the possibility that radiation-induced
mutation could have played a role.

As indicated in Chapter Seven, the Japanese associated the rapid
proliferation of *Candida Albicans* with a radiation-induced fungal
mutation after Hiroshima. In the same period they also noted the
sudden emergence of several previously extremely rare forms of
cancer, such as pancreatic cancer and childhood leukemia. In this
country, Lyme disease may have suddenly become epidemic because of a sudden lethal change in a spirochete that had been
carried by deer and field mice for prior generations without harm
to humans. As related in the previous chapter, an outbreak of Lyme
disease began in the fall of 1975, after huge radiation releases from
the nearby Millstone reactor.

A presentation of the hypothesis that radiation led to mutations
which gave rise to AIDS was made by Ernest Sternglass and Jens
Scheer at the annual Conference of the American Association for
the Advancement of Science on May 29, 1986.[179] The response of
the scientific community to this paper was complete and utter
silence, despite the fact that it provided a potential explanation for
why and where AIDS first emerged. In the words of Professor
Scheer of the University of Bremen, it was as if the paper were
"spurlos versunken" (sunk without a trace).

Here is a summary of this paper, which begins as follows:

*Two of the principal unexplained aspects of the AIDS
epidemic are the timing of the sharp rises beginning in
1980–82 and the initial geographic concentration in
Central Africa, the Caribbean, and the East and West*

Coasts of the U.S. These findings can be explained by the hypothesis that beta irradiation of bone marrow cells by strontium-90 and other bone seeking radioisotopes in the diet during the period of nuclear testing may have led to mutation of an AIDS related indigenous human or animal retrovirus, and also produced a cohort of susceptible individuals whose immune defenses were impaired during intra-uterine development.

Although atomic bomb tests began in 1945, the world-wide increase of strontium-90 in the diet did not take place until after the large hydrogen bomb tests of the mid-1950s, rising most sharply between 1962 and 1963. The greatest increase in AIDS occurred some 18 to 19 years later, or between 1980 and 1982, when the large cohort of potentially immunodeficient infants would have reached maturity and been exposed to sexually transmitted diseases. Thus, the AIDS virus would spread among people of this cohort wherever conditions favored high rates of sexual contact or other efficient means of transmission directly to the bloodstream.

Since 90 percent of fallout comes down with precipitation, this hypothesis could explain why the AIDS epidemic began in areas of high rainfall such as Central Africa and the Caribbean Islands close to the latitude of the Pacific test sites and rose most rapidly in the high rainfall areas of the East and West Coasts of North America, with fewer cases per million persons occurring in drier regions such as North or South Africa or the Central Plains states of the U.S.[180]

Because strontium-90 is transmitted mainly through the diet, it also explains why South-East Asia, although high in rainfall, shows few AIDS cases, since rice and fish have a much lower strontium-90 to calcium ratio than milk, bread, meat, fruit, potatoes, beans and vegetables dominant in U.S., Caribbean and African diets.

The authors cite as examples recent measurements taken in New York in 1982, which indicate that fresh fish and rice contain one tenth the amount of strontium-90 per unit of calcium observed in fresh vegetables, potatoes and dry beans. Thus:

Depending on the type of diet, people living in the same area can have as much as a ten-fold difference in their intake of strontium-90 for every gram of calcium needed for the formation of bone in the developing infant, the young child, and the adolescent before pregnancy.

They further support their hypothesis by referring to a 1957 study published by the United Nations Scientific Committee on the Biological Effects of Atomic Radiation, which found that the highest concentrations of strontium-90 in human bone were in the high rainfall areas of Africa, with measures four times greater than those in much-drier South Africa. They continue:

The crucial role of diet could also explain the puzzling fact that on the island of Trinidad in the West Indies, with equal populations of African and Asian Indian origin and comparable percentages of homosexuals, there were 45 cases of AIDS diagnosed among those of African origin and none among those of East-Indian descent. A detailed study of the dietary differences and strontium-90 concentrations in bone for the two ethnic groups would serve as a valuable test of the present hypothesis.

There is therefore persuasive evidence that fallout from atmospheric bomb testing may have damaged developing immune systems in the early 1960s, and may have also accelerated the mutation of an AIDS-like virus found in indigenous African monkeys that made it more virulent to humans. One of the difficulties in treating the virus is that it appears to be still mutating, resulting in more than one strain.[181] The adverse effect of low-level radiation from ingested radionuclides has been the subject of dozens of laboratory studies, referred to throughout this book. A recent paper, for example, carried the evocative title "Ultraviolet radiation inhibits human natural killer cell activity and lymphocyte proliferation," referring to the effect on the blood cells responsible for the immune response to foreign agents.[182]

Chronic Epstein Barr Virus (CEBV) is another recently virulent disease linked with a weakened immune system.[183] It has been associated with a rare form of cancer known as Burkitt's Lymphoma, first encountered in Uganda in 1965. In equatorial Africa, patients with Burkitt's Lymphoma have the same aberrant immune system response as occurs with CEBV patients. The good news is that CEBV patients respond to several therapeutic and dietary treatments designed to boost the immune system.

The cost of performing a laboratory assay of the strontium-90 content of human bone is about fifty dollars. It would be relatively inexpensive to conduct a study that compares the concentrations of strontium-90 in the bones of individuals who died of AIDS with those of people killed in accidents, controlling for age, place of birth, socio-economic status, dietary patterns, etc. Foundations established to encourage new lines of medical research could easily fund such a study. Sooner or later, younger scientists will be drawn to this investigation, which older members of the scientific and medical establishments have ignored. The fact that humans have no (or poor) immune defenses against CEBV, AIDS, and Lyme disease is another indicator that these are brand-new viruses. Hence mutation (by whatever cause) is a logical suspect.

CONCLUSIONS

CHAPTER 11

IT'S NOT TOO LATE

This book has tried to indicate the potential toll in human lives that low-level radiation from nuclear bomb tests and reactors has exacted over the past four decades. Because of national security concerns going back to the early days of the Cold War, the truth about such losses has been withheld from the American people.

The federal government has recently admitted that radioactive contamination caused a significant loss of life on at least two occasions, and in both cases denied victims the right to sue for damages. After acknowledging that uranium escaped from its Fernald nuclear fabrication facility near Cincinnati, Ohio, the Department of Energy absolved its contractor, National Lead, from any financial liability. The Supreme Court reversed a multi-million dollar award to residents of St. George, Utah, who proved that cancer deaths there were associated with fallout from mishandled bomb tests. In effect, the government has asserted a sovereign right to endanger the lives of its citizens, as if we have been in a state of war all these years.

It is easy to understand how national security was invoked to withhold information about radiation releases from the Savannah River Plant in order to protect tritium supplies critical for producing thermonuclear weapons. But can national security justify concealing an enormous loss of lives?

It is more difficult to explain the role of the scientific community and the media in sustaining this deadly deception. There has been an almost complete absence of serious debate in American scientific and medical journals about the effect of ingested or inhaled fission products on the hormonal and immune systems. A distinction should be made between the nuclear scientists who permitted national security to take precedence over unwanted truths, and the majority of scientists and physicians who have been unaware of the evidence that free radical-induced biological damage may be thousands of times more efficient at low doses of radiation than at high ones.

Despite the warnings of Rachel Carson, Linus Pauling and Andrei Sakharov, there is nothing in the century-long experience with brief exposures to high intensity X-rays and radiation to prepare physicians to understand the distinctly different biochemical mechanisms involved in internal low-level radiation. Once radioactive fission products come down in the rain and enter the food chain, immune systems become vulnerable to free radicals by means quite different from the destruction of DNA by high-level radiation. This was not known by many of the nuclear scientists who developed the atomic bomb, and by biologists concerned with genetic damage.

These scientists would have been hard pressed to envision the perverse nature of the food chain that causes certain ingested fission products to accumulate in much higher concentrations than naturally occurring isotopes. For example, when cows graze over large exposed areas, the radioactive iodine will concentrate in them. When people ingest contaminated milk, water, root vegetables, or fruits, the adverse effects continue to multiply as the radioactive substances concentrate in organs such as the fetal thyroid or the bone marrow of young women prior to pregnancy. Finally, who could have expected the perverse supra-linear nature of the dose response, with lower levels of radiation potentially being hundreds to thousands of times more efficient in producing the free radicals that penetrate and destroy the blood cells of immune systems?

If this knowledge is deemed subversive and is thus excluded from established scientific journals, physicians will never consider the potential effect on the immune systems of their patients. The discoveries of Drs. Sternglass and Petkau, for example, were published in technical European professional journals, and have rarely been considered by physicians. After our Chernobyl findings prompted a *Toronto Globe* reporter to question Dr. Petkau about his work, he was warned against offering further interviews by his employer, Atomic Energy of Canada, Ltd.

Unlike the government and the nuclear industry, physicians have no vested interest in perpetuating nuclear myths on political or economic grounds, and may—indeed must—become increasingly concerned about the effects of free radicals on the human immune system. The most detailed coverage of the Chernobyl findings was carried in the American Medical Association's *Medical News* and its Canadian counterpart in February 1988. Even the weekly medical news section of *The New York Times* has identified free radicals as a "major cause of disease," especially with respect to immune deficiencies.[184] But the *Times* failed to mention Dr. Petkau's crucial discovery that this type of damage is thousands of times greater for protracted exposures to internally deposited fallout than for short exposures to medical X-rays or gamma rays.

It has long been known that the body's immune defenses detect and destroy cells that are out of control, having become malignant. In *Secret Fallout*, Dr. Sternglass offered the following analogy for human society:

> It is the freedom to investigate and communicate important scientific or public health findings quickly and widely—no matter how disturbing or controversial— that is the key element in the protective system needed to alert a society to potentially dangerous developments before they become irreversibly destructive.[185]

This analogy illuminates the crisis now facing both the U.S. and the U.S.S.R. The Chernobyl disaster truly shook the world and may be a final warning that the prospects for continued life on earth are put at risk by nuclear technology. This helps explain Gorbachev's surprising speech before the United Nations in December 1988, in which he cited the environmental threat to the planet as overriding the rationale for the nuclear arms race.

If two American researchers using published monthly data could identify 40,000 lives that appeared to be cut short by the Chernobyl fallout, thousands of miles away from the accident, consider what Soviet and Polish officials must have found in their unpublished mortality records. They could certainly assess the health impact of vast agricultural areas contaminated by fallout.[186]

In June of 1987, Adrian de Wind, President of the Natural Resources Defense Council (NRDC), delivered our Three Mile Island findings, which had just been presented to the staff of the Senate Public Health Committee, to Evgeny Velikhov, vice president of the Soviet Academy of Sciences. At the time, the NRDC was leading an effort to install equipment in the Soviet Union to monitor nuclear bomb tests. On his return from Moscow that summer, de Wind reported that Velikhov found the paper surprising, since "TMI was nothing compared to Chernobyl." He encouraged us to send Velikhov the just-completed paper on the impact of Chernobyl fallout on U.S. mortality. We sent Velikhov the Chernobyl findings, with some degree of trepidation that they would not inform him of anything he did not already know. Velikhov did not respond.

It will take a lot more *glasnost*—in America as well as the Soviet Union—for the true dimensions of the Chernobyl tragedy to be acknowledged. Almost two years after the accident, buried in a small item on gold futures in *Investors Daily*, the report was that:

> *Gold futures firmed Friday, rallying out of two year lows on news that Moscow had ordered the evacuation of 20 more villages due to lingering radiation from the 1986 Chernobyl nuclear accident.*[187]

Reports of the evacuation was not made available to readers of most major newspapers.

The American press also avoided coverage of our embarrassing Chernobyl findings for months until they succumbed to competitive pressures from Japan, Italy, West Germany, Canada, and England, where the story was considered worthy of front page coverage.

The *New Scientist* related how the prestigious British journal *Nature* accepted hundreds of papers from scientists around the world on Chernobyl's effects, but then failed to print them. The would-be authors complained in an open letter, that:

> *We appreciate the journal* Nature *as one of the leading scientific journals in the world. The present situation embarrasses us deeply. We feel that not only the case of our present manuscripts but important questions of principle are involved, e.g., the right of the scientist to know what happens to his unpublished work. If his work is rejected, the author has a chance to improve it and try again. We have now encountered a new, much worse alternative: to be accepted and not published.*[188]

Does the West need its own Chernobyl for its leaders to practice *glasnost* with respect to its own nuclear technologies? At the time of this writing key American nuclear fabrication facilities have been shut down. Not only are workers reluctant to enter the contaminated facilities, but their supervisors also share these fears. Even the corporate contractors who have operated the facilities for decades appear to be having second thoughts, as the facilities exceed their lifetime limits.

Senator John Glenn is among the few U.S. politicians who have declared the need to "tear away the veil of secrecy and self-regulation." In a *New York Times* column on revelations from his hearings regarding the Fernald (Ohio) facility, he stated:

For decades, the Government has violated its own worker health and safety standards and has frequently ordered the private contractor running the facility to ignore state and federal environmental laws. As a result, vast quantities of radioactive and toxic wastes are contaminating offsite drinking water supplies. Residents live with the fear that the plant may have harmed their health and that of their children. Now, adding insult to injury, the government proposes to close the operation, with statements that the severe environmental contamination will be cleaned up—sometime—but just when it does not say.

How could this happen? Secrecy. Back in the 1950s, the production of nuclear weapons material was paramount, and secret.[189]

Thus, for Americans as well as Soviets, openness must replace secrecy. As Senator Glenn concluded, "it will do us little good to protect ourselves from our adversaries if we poison our own people in the process."

Though considerations of life and death have often been outranked by economic considerations in the formation of our nuclear policies, the costs are now too staggering for us to bear. The human losses attending every major release of fission products underscores the truth that thermonuclear weapons cannot be used.

The U.S. General Accounting Office estimates that it may cost $175 billion to clean up and replace the military's nuclear production facilities. Add to this the far more daunting problem of what to do with the radioactive wastes now accumulating in pools at every civilian reactor across the nation. The national nuclear cemeteries for civilian and military high-level radioactive wastes, proposed for Nevada and New Mexico, may be myths in political, technological and economic terms. As physicist Marvin Resnikoff has demonstrated, simply by transporting the millions of curies of deadly materials to repositories nobody wants, sixteen accidents per year could occur, any one on the scale of Three Mile Island.[190]

These huge volumes of nuclear waste may end up staying just where they are. Every reactor may end up being entombed, along with its wastes, as the Soviets have had to treat Chernobyl. But who can guarantee that the wastes will not leak? New technologies are needed to guarantee permanent cooling and containment, to keep the wastes from contaminating underground aquifers, on which future generations will rely for drinking water.

The cost to dispose of all nuclear facilities in the next century will at least be on the same order of magnitude as the cost of constructing them. A rough estimate of four decades of defense expenditures earmarked for nuclear weapons and three decades of federal subsidies to the civilian nuclear industry yields a figure in the range of two trillion dollars. And right now, when one considers the associated drain on scientific manpower and other human resources, it becomes clear why our deficit-ridden economy can no longer compete with the demilitarized economies of Japan and West Germany. A recent *New York Times* editorial traced American economic ills to diminishing productivity growth since 1965, at a cost, ironically, of two trillion dollars.[191]

Former President Reagan was probably right when he said our military capability was second to none. Yet as the economist Benjamin Friedman said of the consequences of Reagan's policies, "for America to earn its international position primarily by military might, as its economic power seeps away, means ultimately that we become a mere policeman, a hired gun . . . Hessians of the twenty-first century."[192]

It may be more than a mere coincidence that U.S. productivity gains have sagged just as the damage to immune systems of the baby-boom generation appeared to emerge among young adults. Consider the implications for U.S. productivity of Ernest Stern-glass's discovery, later supported by two U.S. Navy psychologists, of the adverse impact of bomb-test fallout on Standardized Achievement Test (SAT) scores. In *Secret Fallout*, Sternglass described his reaction in 1975 when reading a *New York Times* article on the puzzling but steady decline in SAT scores since the mid-1960s, generally by no more than two or three points per year until 1975, when they dropped by ten points in a single year:

> *Suddenly, the question flashed through my mind: When were these young people born or in their mother's womb? Most of them were 18 years old when they graduated from high school. What was 18 taken from 1975? It was 1957, the year when the largest amount of radioactive fallout ever measured descended on the U.S. from the highest kilotonnage of nuclear weapons ever detonated in Nevada.[193]*

By 1979, with the help of the educational psychologist Dr. Steven Bell, Sternglass was able to secure state breakdowns of the SAT scores, which indicated that the greatest declines had indeed occurred in states closest to the Nevada Test Site. The greatest decline was registered in the neighboring state of Utah, where the large Mormon population had the lowest rates of cigarette, drug and alcohol consumption in the nation, and traditionally had very high SAT scores.

These findings were presented at the annual meeting of the American Psychological Association in September 1979. There it was predicted that SAT scores would begin to improve again in 1981, 18 years after atmospheric bomb tests stopped in 1963.

Two Navy psychologists investigated whether these findings could throw light on the difficulties new recruits were having in mastering complex weapons technologies. They found that:

> *the state having the largest drop in [SAT] scores from children born during the two year period 1956–1958 was Utah, a fact which is consistent with Utah's proximity to the Nevada Test Site and the general northeastern mo-*

tion of the fallout clouds produced by the Nevada tests, providing very convincing and disquieting evidence closely linking the SAT score decline to the cumulative effects of nuclear fallout.[194]

These "disquieting" findings were largely ignored by the media, as was the fact that SAT scores have risen since 1981, confirming Sternglass's prediction.[195]

After the Chernobyl fallout in the summer of 1986, the German edition of *Psychology Today* reexamined Sternglass's SAT findings, raising concerns about possible mental impairment of German children who would reach their 18th birthday in the year 2004. Yet if a theory's validity lies in its predictive value, then Sternglass's SAT hypothesis would appear to have already passed the test.

What are the potential social costs associated with those members of the baby-boom generation who survived birth in the atmospheric bomb test years, but with physical and mental impairments that may hinder them from playing responsible roles in the work force today? Facing a bleak and poverty-stricken future, these young people would be an increasing burden to society as they swell the ranks of the drug-addicted, the homeless and the overcrowded prison population.

Dr. Charlotte Silverman published research in 1980, which found that children who had radiation treatment for ringworm experienced significant mental deterioration years afterwards.[196] Dr. Silverman summarized her results and similar findings for Israeli children at the Sixth International Congress of Radiation Research in Tokyo as follows:

Several measures of brain function, mental ability and scholastic achievement demonstrate that the irradiated children suffered impairment. These findings are consistent with and extend previous findings of suggestive brain damage from radiation.[197]

The sociologist Dr. R. J. Pellegrini has studied an FBI database of Uniform Crime Reports going back to 1945 and discovered that rates of criminal homicide, forcible rape and aggravated assault doubled in the 1970's as compared with previous decades, just as the baby boomers entered in the age group 15 to 24. Crime rates for those 15 to 34 years of age are now at all time peaks, a fact Dr. Pellegrini attributes to their exposure to radiation from fall-out.[198]

Businesses are spending millions of dollars a year on remedial reading and arithmetic instruction, because many young adults entering the labor force are unqualified for work.[199] To what extent might exposure to bomb-test fallout contribute to this deterioration of abilities, which is most commonly blamed on a breakdown in the American school system?

Finally, consider the explosion in medical care costs in recent decades. While examining the health effects of Chernobyl in Europe, Professor Jens Scheer of the University of Bremen discovered that a West German health insurance company experienced the largest annual cost increases for allergic diseases in more than ten years as a result of increased demand in the months after Chernobyl.[200]

Since 1970 total private and public expenditures on health in the U.S. have been rising at more than twelve percent each year. Projecting this increase into the future results in total expenditures of 2.5 trillion dollars by the year 2000, outranking U.S. expenditures for food and shelter!

A most troubling aspect of the current explosion of medical costs, not yet sufficiently appreciated, is the increasing proportion associated with young persons rather than with the aged. Since 1982, and perhaps for the first time in our history, mortality rates for persons aged 15 to 54 are rising. Thus what should be the most productive sector of the labor force must now deal increasingly with the morbidity and mortality problems associated with AIDS, and other immune deficiency diseases.

Yet there are reasons to be hopeful. In the face of the medical and economic crises we face today, it becomes more and more possible for public agitation, expressed through Congress, to decrease our dependence on nuclear technologies. It is not too late. We have only to look at those (admittedly few) areas of the world that have managed to avoid the "benefits" of the atom to realize that we can still enjoy breathing air and eating food that is relatively free of radioactivity.

In our own country, the states of Wyoming and Montana are examples of such areas. Infant mortality rates that were once far worse than the national average have been improving remarkably since 1970. In 1987 and 1988, infant mortality in Wyoming was 42 percent less, and in Montana 30 percent less than in the U.S. as a whole, rivalling countries like Denmark with among the lowest infant mortality rates in the world.[201] Wyoming and Montana have no nuclear reactors and benefited from the pains the government took to avoid testing bombs in Nevada when winds would have blown fallout toward Salt Lake City and Canada.

A change in nuclear policy would free scientific resources to grapple seriously with the energy crisis, an effort long delayed by the nuclear option. A tiny fraction of the trillions of dollars spent in the past four decades on this deadly technology could realize the promise of improving energy efficiencies, and of developing solar and other less destructive energy-producing technologies. A free science will be necessary to find ways to cure the immune system deficiencies that now plague the world and that may have been partly caused by low-level radiation. And ultimately, more open scientific inquiry will serve the goal of preserving the prospects for continued life on earth.

As I now see it, a curious twist of fate was responsible for my embarking on this controversial voyage of discovery, in such sharp contrast to my professional interests as an economist and statistician.

I was retained by Westinghouse Electric Corporation in 1979 as an expert antitrust statistician in a suit involving the price-fixing of uranium by a so-called "international uranium cartel." My job was to ascertain whether forces of supply and demand could justify a six-fold rise in the price of uranium in the wake of the oil embargo of the early seventies. It became clear to me that uranium shortages could not have been a factor in the rise in the price of uranium. Because of frequent shutdowns, the performance of civilian nuclear reactors in the seventies fell far below expectations, and thus so did their demand for uranium. I was also surprised to find that the monthly reports of reactor operations in the nuclear trade press were more realistic than the contradictory statistics available on the consumption of uranium from the Atomic Energy Commission.

My interest in nuclear problems began with that assignment, especially as I had been appointed to the Science Advisory Board of the Environmental Protection Agency in 1977 as a result of my discovery of a way to estimate regional concentrations of toxic wastes. This, in turn, led me to reflect on possible linkages of environmental problems with effects on public health.

157

It was in this connection that I first read *Secret Fallout,* by
Dr. Ernest Sternglass, which I found to be most provocative and
disturbing. Here was an eminent professor of radiology at the
University of Pittsburgh Medical School, and a former senior physicist at Westinghouse Research Laboratories, who had become
convinced, against his will, that the government was lying about
the mortality impacts of an early accident at the Shippingport
reactor near Pittsburgh. I felt there could be no question about his
honesty as he described the process by which he was forced to
assume the role of an anti-nuclear whistle blower. But I never
dreamed that someday I, too, would be impelled, out of what was
only a modest degree of intellectual curiosity, to confirm his findings and play a similar role.

My career as an expert statistical witness began, quite by chance,
when I was retained by the Department of Justice in 1955 to
prepare statistical exhibits for a small antitrust suit. To the surprise
of all concerned, the suit went all the way to the Supreme Court
and became the famous Brown Shoe Case. In addition to establish-
ing precedents for all postwar antitrust litigations, it established the
role of a statistician as an expert witness in evaluating the significance of estimates of a company's "market share."

As a result of this case, my career as a statistical expert blos-
somed. For the next two decades I participated in more than two
dozen antitrust cases involving many major American companies,
including IBM, Beatrice Foods, Greyhound, Armour, Occidental
Petroleum, R. J. Reynolds, Emerson Electric, North American
Phillips, and Westinghouse. Many of these companies have disappeared in the "merger mania" of the Reagan years so that antitrust
litigation is today merely a historical footnote.

While there may no longer be a demand in antitrust litigation
for the analysis of market share ratios, I believe there is now a great
need for the expert testimony of statisticians in evaluating the
significance of another ratio, which lies at the heart of toxic tort
litigation. The key ratio to study when a population is exposed to
an environmental risk is the "observed" number of deaths in a

given area and period compared with the "expected" number based on national norms. We then seek to ascertain whether the ratio of the two numbers is too great to be attributed to chance. It is this measure of a statistically significant excess death rate change which lies at the heart of the demonstrations in this book of excess deaths following releases of low-level radiation.

My active antitrust practice led me to establish Economic Information Systems Inc., a company which successfully developed large computer databases for the analysis of the market shares of major companies in all industries. After the sale of my company to Control Data Corporation in 1981, I felt free to use my special database expertise to explore environmental problems. In the course of this I came across a remarkable book, *The Next Nuclear Gamble,* by Marvin Resnikoff, published by the Council On Economic Priorities.

From Dr. Resnikoff, I learned that highly radioactive used nuclear fuel assemblies were piling up in refrigerated pools at each reactor, awaiting the time, presumably early in the next century, when a "Great Nuclear Cemetery" would be built to receive them. Resnikoff estimated that the task of transporting the hundreds of millions of tons of radioactive materials away from the reactors was likely to involve an average of sixteen nuclear accidents each year, each of which could spew as much radioactivity into the environment as had the Three Mile Island catastrophe of 1979.

The Resnikoff book was so impressive that I paid a visit to the Council on Economic Priorities (CEP), and accepted an invitation from CEP to continue my environmental research at the Council when my responsibilities to Control Data ended in 1984.

My first research project at CEP affirmed the National Cancer Institute's 1977 findings that cancer mortality at the county level was correlated with concentrations of petrochemical activity. Reasoning that environmental pollutants, initially at least, were highly localized, and remembering that the 1980 Census had for the first time included the five-digit ZIP code area as a geographic unit, I made some calls to the Census Bureau and other federal agencies.

To my surprise, I discovered that at very little cost (and with help from the Freedom of Information Act) I could purchase computer tapes from the Census Bureau, the EPA, the National Cancer Institute, and the National Center for Health Statistics, containing unpublished but politically sensitive information that had cost the government an estimated $40 billion to collect!

This treasure trove of environmental and public health data had never been integrated and analyzed comprehensively. I can only speculate that epidemiologists, mainly employed by state departments of health, rarely investigated the causes of wide variations in local mortality rates, perhaps because of the political consequences of finding a correlation with some local environmental abuse. I believe most environmental epidemiological studies are self-limiting as a result.

For someone with a database background like myself, these unused files were a researcher's paradise. It proved to be relatively easy to find statistically significant differences in geographic mortality rates because the official U.S. mortality databases were based on all death certificates filed in a given year. Even small differences might prove to be significant because of the large numbers involved. However, processing such large databases requires a professional staff of computer programmers and analysts.

So, in 1985 I helped organize a small company called Public Data Access, Inc. (PDA) in the hope that it could eventually serve as the computer research arm of the environmental movement. It was in this context that I first worked with Ben Goldman, who at the time was a project director at CEP. He had discovered the same need to integrate diverse government data sources for a study of the hazardous waste industry, and had ventured into the world of micro-computers to accomplish the task. In fact, the database he developed on a personal computer for *Hazardous Waste Management: Reducing the Risk* (Island Press, 1986) ended up being PDA's first commercial product, available through a computerized information network called Chemical Information Services, Inc. Ben helped me start PDA, and eventually became its president.

This fledgling effort received generous support from environmental foundations and concerned individuals. Even more important was the "sweat equity" contribution of a small group of enthusiastic young analysts, who in time became adept in interfacing large mainframe computers with the increasingly powerful and flexible personal computers, which among their other effects have made possible a decentralizing of data processing and great cost reductions.

Since 1986, PDA has helped produce numerous environmental research studies. *Quality of Life in American Neighborhoods: Affluence, Toxic Waste and Cancer Mortality in Residential Zip Code Areas* (Westview Press, 1986), was a "database publication" with data for each of some 35,000 residential five-digit ZIP Code areas taken from EPA and Census Bureau databases. Although the 1980 Census cost about one billion dollars, the Reagan Administration decided that it would be too expensive to publish results for small areas such as ZIP codes, so this book remains the only public source for such localized information. PDA also produced in 1987 *Toxic Wastes and Race in the United States* for the Commission for Racial Justice, which showed that a significantly disproportionate number of toxic waste facilities were located in African-American neighborhoods. *The Philadelphia Toxics Story*, for the National Campaign Against Toxic Hazards, and *Toxic Waste and Cancer Mortality in Michigan*, for the Public Interest Research Group in Michigan, also were produced by PDA in 1987. In 1988, PDA completed *Mortality and Toxics Along the Mississippi* for Greenpeace USA.

In all of these publications, the focus was on geographic areas with significantly high mortality rates and the extent to which these were correlated with exposures to toxic chemicals. Although frequently the correlations were good, in important cases they were poor. This apparent paradox was only resolved for me by discussions with Dr. Ernest Sternglass, who pointed out that a large culprit in the mortality situation might not be toxic chemicals, but low-level nuclear radiation. He argued that at least some part of the positive correlations we were obtaining between toxics and mortality were due to the overlap of nuclear pollution and toxic pollution in industrial areas, and also that where we were not getting a good correlation it might be because our studies did not contain a nuclear pollution variable.

Dr. Sternglass also realized that our large mortality databases would overcome earlier criticisms of his research that had used small bodies of data to show that low-level radiation had a major impact on mortality. In response to his challenge, I decided to examine recent mortality trends in areas most exposed to nuclear emissions since 1975. In this work I did find small but statistically significant increases in total mortality, cancer mortality and infant mortality, for the period 1975–82 (compared with 1965–69) for "nuclear" as compared to "non-nuclear" areas. Nuclear areas were defined as the 160 counties which either had at least one commercial nuclear power plant or were downwind of such a county.

After months of recalculation and checking, CEP published the nuclear county results in its December 1986 newsletter, entitled "Nuclear Emissions Take Their Toll." In the months that followed, I was disappointed with the complete lack of U.S. news coverage. However, it became a front-page story in Italy in January of 1987. Fabrizio Tonello, who reported on my findings for the Italian news weekly *Il Mondo*, told me that his story played a significant role in generating the 80 percent referendum vote to halt work on Italy's two proposed nuclear reactors. Later that year the Italian cabinet accepted the popular decision, and Italy today has banned the construction of new nuclear reactors.

European environmentalists were not the only people who were aware of my findings, however. I learned the CEP newsletter was making waves in Italy after I received a telephone request from the Italian Atomic Energy Commission for a copy. Only a few hours later, I received a call from a Philadelphia engineering firm involved in the construction of nuclear reactors. My caller seemed somewhat flustered at reaching me directly, as I had no secretary at the Council. I asked whether he wanted a copy of the report. "No," he replied, the question he wanted answered was, "who authorized this report?"

In the U.S., various anti-nuclear groups throughout the country, disturbed by the CEP newsletter, clamored for detailed reports on each reactor. After more detailed examination of our databases, I began to believe that most of the excess mortality I had found may have been associated with a few major accidental nuclear releases, especially the accident at Three Mile Island (TMI).

Concerned Harrisburg residents urged us to present our TMI findings to the Senate Public Health Committee, with a plea for public debate. At this time I first learned that there were 2,500 lawsuits in progress against the local utility, of which 300 had already been settled on condition that no details of the settlement would ever be revealed. All this despite the claim that "no one died at TMI."

I presented our TMI findings to staff members of the Committee twice in the Spring of 1987. Early in 1988, Senator Edward Kennedy, Chair of the Public Health Committee, requested that the National Institutes of Health conduct a study of mortality near nuclear reactors. Senator Kennedy cited reports of high leukemia rates found by epidemiologists of the Harvard School of Public Health near the Pilgrim reactor in 1982–84, which were published in a letter to the British medical journal *The Lancet* on December 5, 1987. *The New York Times* on July 7, 1988, reported that the National Cancer Institute (NCI) had agreed to study cancer deaths among people living near nuclear plants. The *Times* quoted Dr. John Boice, chief of radiation epidemiology at NCI, who said "the study was prompted by a British survey completed last year . . . [that] found a higher incidence of leukemia among children and teenagers living near nuclear plants."

We expect that this book may contradict the NCI findings, promised for 1990. Our hypothesis suggests that many more counties need to be studied than the number proposed by NCI. Dangerous fission products, particularly radioactive iodine and strontium, can be borne by winds and waterways for hundreds of miles and then come down in the rain, contaminating sources of fresh water and milk far from the reactor site itself. These ingested fission products can then wreak serious damage on immune systems. All of this was suggested by our Chernobyl findings.

We began our Chernobyl study after an invitation to present our TMI findings at the European Conference on Chernobyl Radiation in Amsterdam at the end of May 1987. While there, I not only gained much anecdotal knowledge about the impact of Chernobyl radiation in Europe, but also learned that no European nation publishes monthly mortality reports similar to the *Monthly Vital Statistics Report* of the U.S. National Center for Health Statistics. I wondered whether mortality in the U.S. could have been affected by the small percentage of Chernobyl radiation that drifted over in the stratosphere and came down in the May 1986 rains.

We began our research on this question when I returned to New York in June 1987, and were intrigued to find that abnormally high levels of radioactive iodine had been detected by Environmental Protection Agency milk-monitoring stations in almost every state beginning on about May 9th, 1986. We also found that significant mortality increases had occurred in May for which the Chernobyl fallout seemed to be the only plausible explanation.

We revealed our Chernobyl findings in two papers delivered at the First Global Radiation Victims Conference held in New York City in September 1987. As we had come to expect, the U.S. press did not cover the story, but it was front-page news in the Japanese and Canadian press. This was followed by major stories in leading English papers such as the *Independent* and the *Economist*. Finally, almost six months after we presented our findings, *The Wall Street Journal* broke the silence in the U.S. by reporting on our results on February 8, 1988.

A few weeks after the publication of *The Wall Street Journal* story, I received a fascinating letter from Dr. David DeSante, a researcher at the Point Reyes Bird Observatory in California, enclosing an article he had published in an ornithological journal early in 1987. He had recorded a 62 percent drop in the number of newly hatched landbirds during the period from mid-May to mid-August, 1986. The only explanation he could find for the landbird reproductive failure was the radiation fallout from Chernobyl.

Why were birds so affected? Why had AIDS-related deaths in the U.S. doubled in May of 1986? Why were human immune systems so damaged by the Chernobyl fallout? All these were questions we wanted to answer. So we started to investigate the major nuclear releases in the past—from atmospheric nuclear bomb tests in the fifties and the sixties to civilian and military reactor accidents such as those at Savannah River, Millstone, and recurrent releases at TMI and Peach Bottom reactors.

Beginning in 1945, the superpowers released massive quantities of fission products into the biosphere from above- and below-ground explosions of nuclear devices with yields of about 600,000 kilotons, according to estimates by the National Resources Defense Council. This was the equivalent of 40,000 Hiroshima-sized bombs. Emissions from nuclear reactors, including major accidents such as Three Mile Island and Chernobyl, added to the total, much of which was composed of long-lived radioactive isotopes that will remain in the stratosphere for millennia.

In the Fall of 1988, the Senate Government Operations Committee, under the leadership of Senator John Glenn, held hearings about a series of accidents and safety problems at military nuclear facilities operated by the Department of Energy. It turned out that crucial information about some of these accidents had been withheld from the public and the Congress for as long as 25 years.

We immediately realized that our mortality database, with its ability to yield calculations of excess mortality at any place and at any particular time, could be brought to bear to illuminate the health consequences of these incidents. In late 1988, we joined efforts with the Commission for Racial Justice of the United Church of Christ (CRJ) in creating the Radiation and Public Health Project (RPHP). The project grew out of a long-standing relationship between PDA and CRJ. Since 1982, CRJ has investigated the presence of toxic substances in residential communities across the country, and has challenged the disproportionate impact on racial and ethnic neighborhoods. This pursuit lead CRJ to engage PDA in 1986 to prepare the ground-breaking study *Toxic Wastes and Race in the United States,* which was the first comprehensive empirical study of race and toxics in the U.S., and which has been instrumental in influencing the Centers for Disease Control to undertake epidemiologic studies in this area. The function of RPHP is to extend our studies of the health effects of nuclear releases, and to stimulate public debate on these controversial issues.

In this book we have drawn heavily on the databases created by Public Data Access, Inc., described in the methodological appendix. I regard these databases as our major achievement, for they represent a wonderful public resource, and a tribute to an important democratic tradition of open access to sensitive information. These databases enable us now to examine, in great detail, clusters of excess deaths from any cause, in any county or groups of counties, anywhere in the U.S., at any time since 1968. While we have focused attention in this book on the long neglected factor of low-level radiation, we fully recognize that all environmental abuses should be analyzed in the same spirit of open public inquiry.

Statistical epidemiology, the study of the distribution and determinants of disease among human populations, goes back many years. Indeed, as John Allen Paulos notes in his book *Innumeracy*, probability theory began in the seventeenth century with gambling problems and "statistics began in the same century with the compilation of mortuary tables, and something of its origins stick to it as well."[202]

Epidemiologists emphasize the fact that statistical correlation cannot prove causality. Indeed, they have coined the phrase "ecological fallacy" to indicate cases where people have mistakenly drawn the conclusion that A caused B from parallel movements of factors A and B. Every epidemiology textbook is full of examples of how one can make this erroneous conclusion.

Common sense tells us the same thing. Any baseball fan knows that if a team won 15 out of 18 games (factor B) in August, this probably has less to do with the fact that the temperature on the winning days was over 90 degrees (factor A), than with a star pitcher or hitter returning to the lineup (factor C). We would reach this conclusion even if 90-degree days correlate better with winning, because the star player may have been in the lineup on losing days too. Thus, if there is a plausible theoretical mechanism by which factor A could cause factor B, then their correlation is much more plausible than one with some factor C for which there is no obvious causal connection to B.

Statistical correlation has always been used to help identify potential causal factors. The search for statistical correlation is not meant to replace causality studies, but rather to provide clues where to look. For example, if one finds an outbreak of sickness in a local community, the epidemiologist would look at many factors, trying to identify which correlates with the illness. If it turns out that most of the people in the community who became sick had eaten dinner at Tom's Restaurant within the last week, it would be a reasonable first step to examine the restaurant or something in it for a causal factor. Follow-up studies scrutinizing the restaurant's cleanliness, food packaging, etc., could ultimately prove that something at Tom's caused the outbreak.

However, an examination of the people who became ill in that community may also show a statistical correlation between the illness and wearing the color red. The sensible epidemiologist, faced with two factors to investigate, factor A (eating at Tom's restaurant), or factor C (wearing red), would undoubtedly put his/ her resources into investigating factor A. Common sense suggests that eating at a restaurant is a much more plausible candidate for causing the illness than is wearing red; though, a curious epidemiologist might also check out the possibility that people were wearing clothes dyed red with a harmful chemical.

In this book, the causal hypothesis linking factor A (low-level radiation) with factor B (excess deaths) is the "Petkau effect," discussed in detail below. The Petkau effect offers a plausible explanation that suggests low-level radiation may in fact cause excess mortality, and this hypothesis is supported by the statistically significant correlations.

This brings us to the important notion of "excess death" used throughout the book.[203] Epidemiologists use the concept of excess death to show that certain geographic areas and demographic groups suffer from unexpectedly high mortality rates. Excess deaths may be roughly defined as the difference between the number of deaths observed in a given population and the expected number. It is relatively easy to measure the observed number of deaths using government tabulations of death certificates. The more difficult question is: how do we know how many deaths to expect?

The most common method epidemiologists use to estimate expected deaths is to compare the population of concern, for example, residents in counties surrounding the Savannah River nuclear plant, with a much larger population, such as all residents of the United States. The basic idea is that the much larger U.S. population experiences a "normal" or average rate of mortality that can be used as a yardstick. Thus, the observed mortality among the smaller population, be it a locality, a particular age cohort, or other grouping, is tested to determine if it is significantly different from the national norm. The smaller group is often called a "sample" taken from the "universe" of the U.S. as whole.

To make this comparison, epidemiologists first "standardize" the population of concern to rule out the influence of peculiarities in age, gender, and racial composition. These three characteristics are most commonly standardized, partly because such data are systematically collected on death certificates. For example, if a county has a much higher proportion of older people, then it is natural to expect a higher mortality rate. Similarly, since women tend to live longer than men, if a county has an unusually high proportion of men, this too might account for a higher mortality rates. In addition to classifying deaths according to age and sex, the government differentiates between "whites" and "nonwhites," with the latter including a wide mix of African Americans, Spanish Americans, Asian Americans, Native Americans, and so on. Although "nonwhites" is thus a very imprecise category, it generally has been observed to have significantly higher rates of mortality than the category "whites."

Epidemiologists deal with these variations by first dividing the population of concern into the different age, gender, and race groups, and then calculating the expected mortality rates of each age-sex-race group based on the corresponding national rate. The calculation of theoretically expected mortality can then be compared with the mortality rate that is actually observed in each group to yield the difference. Thus excess death is defined as the number of observed deaths that are significantly higher than expected for each race-sex-age group based on their corresponding national average; although, sometimes, only adjustments for age are used.

It is important to note one flaw in the conventional standardization technique used by government epidemiologists, and employed here. By standardizing for the vague racial categories of white and nonwhite, this technique underestimates the excess deaths suffered by people of color. Rather than considering the cause of higher mortality among nonwhites, this technique simply defines it as expected.

There are clear biological and behavioral explanations for expecting higher mortality among certain sex and age groups; the same cannot be said for the multi-racial grouping called nonwhite. A person over eighty is more likely to die than a young adult. Females are more likely to get breast cancer than males, and males are more likely to have a heart attack than females. The majority of differences in mortality among racial and ethnic groups, on the other hand, are caused by environmental factors, including living conditions, diet, pollution, etc. Only a very few diseases have been genetically linked to certain racial and ethnic groups, Tay-Sachs among Jews, for example, and sickle-cell anemia among African Americans.

Defining higher mortality among nonwhites as expected is thus similar to saying that society expects nonwhites to be exposed to unhealthy environmental factors. Future research should adjust for this distortion; however, this was not done here.

What elevates an excess death to the level of being "statistically significant?" In rough terms, an event is statistically significant when it is "improbable" that it would be observed in the real world if merely the laws of chance were operating. Epidemiologists seek fluctuations in mortality that exceed the limits of chance variation. They do this by determining the "improbability" of the difference between observed and expected mortality. A difference so great that it is improbable that it results from chance means the observed mortality increase or excess is "significant."

Throughout this book, an increase in mortality is characterized as significant if the probability that it could be due to chance is less than one out of 100 (a "P value" of less than 0.01). This judgement can be made precisely, because variations in mortality conform to the bell-shaped "normal" curve. Because the statistical demonstrations are intended to develop hypotheses rather than to prove their validity definitively, significant divergences have been identified occasionally with P values of less than 0.05 (less than five percent probability of a chance result). Both confidence levels are commonly used by statisticians. Any individual case that passes a significance test may still reflect a random variation. But the cumulative significance of the five sets of correlations between low-level radiation and increased mortality, considered in Chapters Two, Four, Five, Eight and Nine, means that the likelihood that they are all chance occurrences is remote.

Imagine repeatedly tossing one hundred coins and recording the percentage of heads that turns up in each repetition. The following sequence could result: 51 percent heads, 48 percent heads, 50 percent heads, 47 percent heads, 52 percent heads, 50 percent heads, etc. If we keep tossing the one hundred coins one thousand times and then plot the number of times each percentage appears, we would generate the bell-shaped normal curve, and 50 percent heads would be the most frequently recorded, or "mean," result.

Statistical theory enables us to determine that the "standard deviation" of this distribution is plus or minus five percent heads. This means that about two-thirds of all results are expected to fall within one standard deviation on either side of the mean result (from 45 or 55 percent heads). About 95 percent of all possible results are expected to fall within the interval between 40 and 60 percent heads, or two standard deviations on either side of the mean. Most statisticians would regard the remaining possible outcomes (i.e., less than 40 or more than 60 percent heads) as highly improbable and thus statistically significant. The following table

indicates the P value, or degree of statistical improbability associated with three increasingly improbable outcomes:

PERCENT HEADS	STANDARD DEVIATIONS	P VALUE
65%	3	0.001
70%	4	0.0001
75%	5	0.000001

When an increase in mortality has a P value of less than 0.001, that
is, less than one out of one thousand, this is equivalent to the highly
improbable act of tossing one hundred coins and getting 65 heads
and 35 tails.

The formula used in this book for computing the significance of
mortality phenomena is the standard one described by the National Center for Health Statistics in the annual volumes of the
Vital Statistics of the United States:

$$(O - E) / SQRT ((O^2 + E^2) / N)$$

where O = observed mortality rate; E = expected mortality rate;
SQRT = the square route of; and N = observed number of deaths.
Expected rates are calculated as a function of the original observed
rate multiplied by the change in the U.S. rate. This formula is
based on a Poisson distribution, which is appropriate for statistically rare events such as mortality. The formula yields that number
of standard deviations by which the observed rate differs from the
expected rate. This value can be converted to a probability estimate
with a table of the area under the normal curve, which can be
found in the back of any statistics textbook.

Since we calculated that it is highly unlikely that the excess
deaths found in the case studies were due to chance, what could
have caused them? The hypothesis proposed here is that they were
caused by a biochemical mechanism whereby ingested fission
products promote the formation of "free radicals" that damage the
immune system. This mechanism was discovered in 1972 by

Abram Petkau.[204] The statistical tests in this book demonstrate that there were highly significant events among large human populations, each of which requires a reasonable explanation. The Petkau effect is a plausible biochemical mechanism (though significance tests cannot prove it was a cause), and thus must be considered.

Dr. Abram Petkau is a Canadian physician and biophysicist who until recently managed the Medical Biophysics Branch of the Whiteshell Nuclear Research Establishment, located in Pinawa, Manitoba. While studying the action of radiation on cell membranes in 1971, Dr. Petkau conducted an experiment never done before. He added a small amount of radioactive sodium-22 to water containing model lipid membranes extracted from fresh beef brain. To his surprise, the membranes burst from exposure to just one "rad" (a measure of the amount of radiation absorbed) over a long period of time. Conversely, Dr. Petkau had previously found that 3,500 rads were required to break the cell membrane when X-rays were applied for only a few minutes. He concluded that the longer the exposure, the smaller the dose needed to damage cells.

After several more experiments, he discovered the cause of this surprising effect from low-level radiation. The irradiation process was liberating electrons, which were then captured by the dissolved oxygen in the water, forming a toxic negative ion known as a free-radical molecule. The negatively charged free-radical molecule is attracted to the electrically polarized cell membrane. This causes a chemical chain reaction that dissolves the lipid molecules, which are the principal structural components of all membranes in cells. The wounded and leaking cell, if unable to repair the damage, soon dies. If the free radicals are formed near the genetic material of the cell nucleus, the damaged cell may survive, but in mutated form. Subsequent research by Dr. Petkau and other scientists ultimately demonstrated that this process occurs even at background radiation levels.[205] At high levels of radiation, Petkau found less cellular damage from free-radical production per unit of energy absorbed than at low levels of radiation.

Free radicals are so dangerous to living systems because they form in water, and water comprises eighty percent of a cell. Free radicals not only destroy healthy cells, but also affect normal cell function in a way believed to speed the aging process.

Nature has provided some protection from free radicals, probably because they are normally produced by the oxygen metabolism within the cell. The protector, superoxide dismutase, quenches the chain reaction.[206]

It is now believed that superoxide dismutase is found in all cells which use oxygen in their life processes. For example, human tissues that contain naturally high levels of superoxide dismutase, such as the brain, liver, thyroid, and pituitary, are more resistant to the effects of radiation than tissues low in superoxide dismutase content, such as the spleen and bone marrow. Apparently this enzyme evolved to protect biological systems from superoxide, or free-radical, damage caused by ultraviolet light, background radiation, and the result of normal energy production in the cell. However, radiation which is produced by fission products and ingested through the food chain, or applied externally, can produce more free radicals than the body can deactivate (or "dismutate"), resulting in gross damage that may be irreparable. Furthermore, Dr. Petkau and others have found that only ten to twenty millirads will destroy a cell membrane, in the absence of the protective superoxide dismutase.

The free-radical reaction can be quenched in another way. At higher intensities of radiation, the free-radicals become so concentrated that they tend to deactivate each other. If this were not so, medical X-rays would cause far greater biological damage than they do. A simple analogy, first used by Dr. Sternglass, can explain this phenomenon. Think of the free radicals as individuals in a crowded room. A fire starts and everyone tries to get out at the same time. As a result, everyone bumps into each other and very few escape. If only a few people are in the room when the fire occurs, however, everyone leaves easily through the door. The rate of escape is very high, and therefore, efficient.

Chronic exposure to low-level radiation produces only a few free radicals at a time. These can reach and penetrate the membranes of blood cells with great efficiency, thus damaging the integrity of the entire immune system although very little radiation has been absorbed. In contrast, short, intense exposures to radiation, as with medical X-rays, form so many free radicals that they bump into each other and become harmless ordinary oxygen molecules. Short exposures thus produce much less membrane damage than the same dose given slowly over a period of days, months, or years.

More recently, Charles Waldren and co-researchers have found that when a single human chromosome is placed in a hybrid cell and irradiated, the ionizing radiation produces mutations much more efficiently at low than at high doses, as in the case of cell membrane damage.[207] They found that very low levels of ionizing radiation produce mutations two hundred times more efficiently than the conventional method of using high dose rates, or brief bursts from X-ray machines. They found that the dose-response curve exhibits a downward concavity (logarithmic or supra-linear relationship) in mammalian cells, so that the mutational efficiency of X-radiation is maximal at low doses, exactly as was found by Petkau for free-radical mediated biological damage. Thus, their findings contradict the conventional scientific dogma that the dose-response curve is linear, and that a straight line can be used to estimate low-dose effects from studies of high doses.

A protracted exposure to ingested beta emitters can be one thousand times more harmful to cell membranes than a brief external exposure to X-rays, because DNA repairs itself relatively efficiently after an X-ray hit compared to the damage caused by oxygen free-radicals at very low doses.[208] This type of exposure may thus account for the jump observed in mortality immediately after nuclear plant accidents, or after fallout from atmospheric bomb tests.

Strontium-90 is chemically similar to calcium and, therefore, concentrates in the bone of the developing infant, child, and adolescent. Once in the bone, strontium-90 irradiates the marrow where the cells of the immune system originate at a low rate over a period of many years. As first discovered by Dr. Stokke and his co-workers at the Oslo Cancer Hospital in 1968, extremely small doses of only ten to twenty millirads can produce visible damage to the blood forming cells of the bone marrow, probably via the production of free-radical oxygen.[209] This can lead to the development of bone cancer, leukemia and other malignant neoplasms both directly by damaging the genes, and indirectly by lowering the ability of the immune system to detect and destroy cancer cells.[210]

A peak accumulation of strontium-90 in the body after two or three years could explain the delayed peaks in total mortality as observed after the Savannah River Plant accidents described in Chapter Four. This accumulation results from the combination of growing uptake and slow excretion, and the consequent mortality primarily involves deaths from heart diseases, as well as from cancers and other causes. Free-radical oxygen, produced most efficiently by internal beta emitters such as strontium-90, may be a factor in coronary heart disease as well as cancer. The theory is that the free radicals oxidize the low-density cholesterol and cause it to become more readily deposited in arteries, thus blocking the flow of blood and inducing heart attacks.[211]

Recent medical research from across the country has provided new evidence linking cancer to impaired immune systems.[212] The studies have focused on transplant patients, who as a group suffer from extremely high rates of a variety of cancers. Their cancers diminished rapidly when the doses of immunosuppressive drugs were reduced. (Such drugs are given to stop normal immune systems from rejecting the transplanted organs.) The researchers suspect two types of cells in the immune system of playing major roles in this phenomenon: natural killer cells and cytotoxic T-cells. They found evidence that during immunosuppression, these cells were more depleted among the transplant patients who developed skin cancers than among those who did not.[213] Earlier research published in 1977 demonstrated that bone-seeking isotopes such as strontium-89 and strontium-90 deactivated precisely such natural killer cells in laboratory mice.[214] These new findings linking cancer to immunodeficiencies, combined with the earlier findings of Pet-kau and others of higher-than-expected cell damage from low radiation doses, point to a possible explanation for the rapid increases in mortality rates after low-level radiation releases.

The correlations of health effects with low-level radiation that are discussed throughout this book may thus be caused indirectly by chronic low-level exposures to ingested radiation through hormonal and immune system damage from free radicals. Low levels of strontium-90 and iodine-131 ingested in food, milk, and water, and breathed in air, may damage the ability of the body to detect and destroy infected or malignant cells. Such damage may occur even if radiation is present at concentrations far below existing standards. These standards were set on the basis of a quite different biological mechanism: cancer cell production caused by the direct impact on genes of high doses of external radiation.

Just before this book went to print, the National Academy of Science's Committee on the Biological Effects of Ionizing Radiation released a new report that bears directly on our principal findings.[215] The Committee's extensive review of the latest scientific literature, known as the "BEIR V" report, concludes that cancer and leukemia risks for the survivors of Hiroshima and Nagasaki have been underestimated by factors of three to four, due to faulty dose estimates and insufficient follow-up study of the survivors. Moreover, BEIR V found that risks from diagnostic X-rays may have been underestimated by an additional factor of two, because they were based on extrapolations of exposures to short bursts of high-energy gamma rays from bomb explosions, which were found to be less effective biologically than X-rays.[216]

The BEIR V report cites numerous studies showing increases in leukemia and cancer rates from very low doses of fallout from weapons testing and nuclear plant accidents. As with diagnostic X-rays, these increases were far above those expected from the studies of bomb survivors, further supporting the principal findings of our book. The report suggests that, "although such studies do not provide sufficient statistical precision to contribute to the risk estimation procedure per se, *they do raise legitimate questions about the validity of the currently accepted estimates* [emphasis added]."[217]

The report goes on to pinpoint what we believe is the basic problem: "the discrepancies between estimates based on high-dose studies and observations made in some low-dose studies could . . . arise from problems of extrapolation."[218] These extrapolations may have led to underestimates at low doses, because they assumed the dose-response curve was linear or quadratic, rather than supralinear (which rises rapidly at low doses and levels off at high doses).

A supralinear dose-response curve is suggested by the so-called "Petkau effect" (discussed in our methodological appendix), which involves tumor promotion from free radicals created by repeated exposures at low dose-rates. Indeed, the BEIR V report explicitly refers to the tumor-promoting effect of free radicals observed in laboratory studies of cells, and illustrates how such promoting agents can dramatically change the shape of the dose-response curves so as to increase the effect of carcinogens at the lowest doses.[219] As a result of risk estimates based on mistaken extrapolations, government standards for environmental releases of radioactivity from nuclear facilities may be 100 to 1000 times too high, especially for infants.

The BEIR V findings of greatest concern for the long run may be the effects of low radiation doses on the physical and mental development of the newborn. Detailed studies of infants who were *in utero* at the time of the bomb detonations in Hiroshima and Nagasaki found a much greater risk of severe mental retardation than previously believed.[220] Moreover, the new studies found that intelligence test scores and school performance of children exposed *in utero* were also significantly affected in relation to the degree of exposure.[221] New studies of children whose heads and necks were irradiated for therapeutic purposes in their early childhood also found behavioral impairment as well as poorer school performance. For example, a study by an Israeli group found irradiated children scored poorly on aptitude, intelligence, and psychological tests, often dropped out of school or entered mental hospitals for neuropsychiatric diseases, and had higher rates of mental retardation.[222]

These results, combined with the new findings of errors in dosimetry, the differences in types of radiation and exposure, and the erroneous assumptions about the shape of the dose-response curve, independently support the correlations of fallout levels with SAT scores in the U.S., which, along with their grave implications for attendant social problems, are discussed in Chapter Eleven. The new evidence led to the following recommendation in the BEIR V report:

> *The dose-dependent increase in the frequency of mental retardation in prenatally irradiated A-bomb survivors implies the possibility of higher risks to the embryo from low-level radiation than have been suspected heretofore. It is important that appropriate epidemiological and experimental research be conducted to advance our understanding of these effects and their dose-effect relationship.[223]*

As well as the need for more research, these findings indicate the need to take immediate steps to reduce considerably the permissible levels of radioactive isotopes in our milk and diet.

The Delaney Clause of the Food, Drug, and Cosmetic Act prohibits any addition to food of substances known to be carcinogenic in man or animal.[224] Yet new studies reviewed in the BEIR V report indicate that radioactive isotopes added to milk and other food by bomb-test fallout are associated with significant increases in leukemia rates in the U.S. BEIR V describes the findings as follows:

> *Leukemia death rates (for all ages and all cell-types) peaked in the decade 1960–1969 and were consistently highest in states with high strontium-90 levels in the diet, milk, and bones (based on surveys by the Public Health Services from 1957 to 1970) and lowest in states with low strontium-90 levels.[225]*

These effects were observed despite the fact that dose rates were well below allowable limits: the estimated total dose over many years of weapons testing was only 400 millirads, compared with a legally permitted maximum individual dose of 500 millirads per year.

The BEIR V report also cites a new large-scale British study by Dr. Alice Stewart and her associates demonstrating that extremely small radiation doses in the environment are capable of affecting the future health of individuals exposed as fetuses.[226] Dr. Stewart had established with earlier research that childhood cancers and leukemias were associated with exposures to diagnostic X-rays during pregnancy. In the latest study, her group discovered a direct correlation of childhood cancers and leukemias with background levels of gamma radiation from natural and man-made sources in England, Wales and Scotland. The cumulative outdoor doses due to this source during fetal life varied between only ten and 40 millirads, with an average of 22 millirads. After correcting for a series of socioeconomic, medical and demographic factors, the researchers found that the effect on fetuses of radioactivity on the ground was more than three times greater than that of diagnostic X-rays.

These findings, based on the follow-up of some 16 million women over as long a period as 36 years, support the conclusion of Dr. Stewart and her colleagues that natural and man-made background radiation may account for the majority of childhood cancers and leukemias in our society today. A total background dose (including cosmic rays and internal sources within the body) of only 150 millirads before birth appears to double the risk of a child dying of cancer or leukemia before age 15. This represents an increased risk of 0.6 percent per millirad, which is many thousands times greater than the 0.8 percent increased risk per 10,000 millirads derived by BEIR V for adults based on exposure to high-energy gamma rays at Hiroshima and Nagasaki.[227]

Dr. Stewart's findings would strongly indicate that the standards set for exposure of adults to low-level radiation may be thousands of times too high for the developing fetus. Her work is based on the Oxford Survey of Childhood Cancers which covers 22,351 cases, a far larger universe than that of the Hiroshima-Nagasaki survivors. Moreover, it uses far superior dosimetry: National Radiological Protection Board measurements of background gamma radiation levels produced by radioactivity on the ground for every ten-kilometer square area in England, Wales and Scotland. It is unfortunate that BEIR V did not quantify this enormous difference between the sensitivity of the developing fetus to low-level radiation and that of the adult.

Numerous other studies, many from England, have examined the effects of low-level man-made environmental radiation on children.[228] One study examined excess leukemia rates among children near the Windscale (Sellafield) nuclear reactors and reprocessing plant on the Irish Sea, near the Scottish border.[229] Another examined children under five years old living within ten kilometers of one or more British nuclear plants, and another looked at childhood leukemia cases around four nuclear facilities in western Scotland.[230] Using an automated technique for locating unusual clusters of cancers, one study identified Seascale, which is near Sellafield, as an area in the Northern and Northwestern regions of England with unusually high mortality from acute lymphoblastic leukemia in children.[231]

The BEIR V report cites many other recent epidemiological studies that also support our findings of much greater-than-expected effects from environmental radiation on adults as well as infants and children. It cites studies of rises in cancer and leukemia among the residents downwind from the Nevada Test Site, studies of participants in American and British nuclear weapons tests, where again leukemia deaths were found to have occurred at rates significantly above those normally expected, despite the very small external gamma radiation doses.[232]

According to BEIR V, a comprehensive survey of cancer incidence and mortality near nuclear installations in England—carried out by the United Kingdom Office of Population Censuses and Surveys—found "significant overall excesses of cancer mortality due to lymphoid, leukemia and brain cancer in children and due to liver cancer, lung cancer, Hodgkin's disease, all lymphomas, unspecified brain and central nervous system tumors, and all malignancies in adults."[233] It is interesting to note that the study found cancer rates did not diminish consistently with distance from the plants. This finding could be explained by our hypothesis that contaminated milk and food produced in rural areas near nuclear plants is frequently transported to the large urban centers, so that there often is not a simple correlation with proximity. Furthermore, a logarithmic type of dose-response, which is quite flat above the smallest doses, would tend to mask any dependence upon distance.

The BEIR V report also cites a study of excess leukemia and other cancers of the blood-forming system in five towns near the Pilgrim nuclear reactor in Massachusetts.[234] This reactor had a series of large releases, culminating in 1982 to 1983, due to a faulty radioactive waste treatment system. Although these were among the worst releases in the history of U.S. commercial nuclear power, their seriousness was kept secret at the time. Sharp rises in Massachusetts' monthly infant mortality rates during the summer of 1982 led to our discovery of large spurious "negative" readings of radioactivity in New England's milk (described in Chapter Six and illustrated in Figure 6–7).

Thus, the BEIR V report's many citations of rising rates of mental retardation, leukemia, and mortality associated with nuclear plants and bomb-test fallout further support our hypothesis that the risks from small doses of environmental radiation have been severely underestimated by government agencies. We suggest in Chapter Six that this tendency may even have led to outright falsification of data.

[1]Brian Jacobs, "The politics of radiation: when public health and the nuclear industry collide." *Greenpeace,* July–August, 1988, p. 7.

[2]*Seattle Times,* Thursday, August 21, 1986.

[3]U.S. Environmental Protection Agency, *Environmental Radiation Report* (EPA 520/5-87-004 Revised Edition), No. 46, September 1986, Tables 9.1 and 15. Iodine-131 is so radioactive that it is measured in terms of picocuries per liter. One curie of iodine-131 is equal to one trillion picocuries.

[4]*Monthly Vital Statistics Report,* September, 1986, Table 3.

[5]The national rise was from 9.6 infant deaths per 1,000 live births in June 1985 to 10.7 in June 1986. See *Monthly Vital Statistics Report,* Vol. 36, No. 6, September 11, 1987, p. 1, and *Monthly Vital Statistics Report,* Vol. 35, No. 6, September 15, 1986, p. 1.

[6]Average peak iodine-131 levels in May 1986 in each state were multiplied by population to derive weighted averages for each of the nine Census regions.

[7]Abram Petkau, "Radiation carcinogenesis from a membrane perspective," *Acta Physiologica Scandinavia, Supplement,* 1980, 492:81–90.

[8]Jay M. Gould and Ernest J. Sternglass, "Low-level radiation and mortality," *Chemtech,* January 1989, pp. 18–21, published by the American Chemical Society, Washington, D.C.

[9]Gunther Luning, Jens Scheer, Michael Schmidt, Heiko Ziggel, "Early infant mortality in West Germany before and after Chernobyl," *The Lancet,* November 4, 1989, pp. 1081–1083.

185

[10]Kate Millpointer telephone interview with Jens Scheer, July 11, 1988.

[11]Based on a national survey, the National Center for Health Statistics determined that in 1986 the incidence of acute medical conditions averaged 189.5 per hundred households, compared with 183.1 for 1985, and 180.8 for 1987. See U.S. Department of Health and Human Services, *Health, United States, 1989*, March, 1989, Table 49, p. 94.

[12]*The Wall Street Journal*, February 8, 1988, p. 6.

[13]*Ibid.*

[14]*Seattle Times*, September 29, 1987.

[15]*Toronto Globe*, July 2, 1988.

[16]*Medical News*, February 26, 1988.

[17]Letter from David F. DeSante, Point Reyes Bird Observatory, to Dr. Gould, February 13, 1988.

[18]The drop was nearly 10 standard errors from the mean of the preceding 10 years.

[19]Petkau (1980). (Cf. 6)

[20]The lethal effect of low-level radiation on plant life is the subject of a remarkable book called *The Petkau Effect* by a Swiss engineer, Ralph Graueb, which is available in French, German and Italian but not English. This book offers a useful summary of the significance of Petkau's discoveries, and of findings of German and Swiss biologists that the recent death of the European forests has been accelerated in areas closest to nuclear facilities. This suggests that the effects of "acid rain" may be aggravated by the interaction of man-made radioactivity with other industrial emissions.

[21]Kate Millpointer interview with David F. DeSante, July 5, 1988.

[22]David F. DeSante and Geoffrey R. Geupel, "Landbird productivity in central coastal California: the relationship to annual rainfall and a reproductive failure in 1986," *The Condor*, 89:636–653.

[23]The quantities of pesticides used, and number of acres sprayed for each agricultural product are prepared quarterly by the California Department of Food and Agriculture.

[24]Kate Millpointer interview with Donald L. Dahlsten, July 14, 1988.

[25]In Blodgett Forest during 1986, 14 of 33 nests failed (42 percent). A high rate of nest failure occurred in 1977, with 13 of 32 failed, but

five of these were due to predators. In 1986, two nests failed due to predation, leaving 12 which failed for unknown reasons, or 36 percent. 98 out of 236 eggs died (41 percent). 57 deaths would be expected, based on a 15-year mean of 24 percent mortality. During no year other than 1986 did observed deaths deviate significantly from the expected number. *Ibid.*

[26]Kate Millpointer interview with C. J. Ralph, September 6, 1988.

[27]Kate Millpointer interview with David F. DeSante, August 8, 1988.

[28]*Ibid.*

[29]With data from a study known as "The Coastal Scrub Avian Ecology Program," which determined the ages of individual birds by examining their skulls, DeSante was able to follow the lives of individual birds of three species in the coastal scrub habitat at Palomarin. He determined a mean survivorship, that is, the mean number of birds that exist in one year and are still alive into the next year. He classified the adult birds into three groups: one-year old birds; middle-aged birds (two to three years); and old birds (four years or older).

[30]Kate Millpointer interview with David F. DeSante, February 15, 1989.

[31]*Ibid.*

[32]*Ibid.*

[33]No longer with PRBO, DeSante is currently conducting independent studies of bird populations, and plans to publish an ornithological journal that approaches the study of bird populations from a global perspective, examining low-level radiation, acid-rain, the greenhouse effect, deforestation of tropical rain forests, plastic pollutants and oil slicks in the ocean, and other global problems affecting birds.

[34]*The New York Times,* October 1, 1988.

[35]*The New York Times,* October 20, 1988.

[36]*The New York Times,* October 19, 1988.

[37]Oak Ridge National Laboratory, *Integrated Data Base for 1988: Spent Fuel and Radioactive Waste Inventories, Projections, and Characteristics,* DOE/RW-0006 Rev. 4, Washington, DC: USDOE, September 1988.

[38]Throughout this chapter, the "Southeast" refers to the two regions

defined by the Bureau of the Census as the South Atlantic (Delaware, District of Columbia, Florida, Georgia, Maryland, North Carolina, South Carolina, Virginia, and West Virginia) and the East South Central region of the country (Alabama, Kentucky, Mississippi, and Tennessee).

[39]Measurements of total beta radiation deposited by precipitation (measured in nanocuries per square meter (nCi/m^2) were taken at a sampling station in Columbia, South Carolina, as well as stations elsewhere in the Southeast and rest of the U.S. The South Carolina readings rose from 1 nCi/m^2 in December 1969 to 6 nCi/m^2 in December 1970; the corresponding rise in the Southeast was from 11 to 24, and in the U.S. from 5 to 9. All radiation data are from successive monthly issues of U.S. Department of Health, Education, and Welfare, *Radiological Health Data* (or *Radiation Data and Reports*), during the years 1965 through 1973, unless otherwise indicated.

[40]The probability is less than one in 10,000 that this was a chance variation above the U.S. trend from February 1969 to November 1970.

[41]The average reading of total beta in the rain was 4 nCi/m^2 for the Northeast, 4.25 for the West, and 0.304 for the Midwest.

[42]The maximum concentrations were 15 picocuries per cubic meter (pCi/m^3) in North Carolina, 6 in South Carolina, and 5 in Alabama.

[43]The probability is less than 1 percent that the July, 1971 reading was a chance variation above the 8-year trend. Measurements of strontium-90 in milk (measured in picocuries per liter, pCi/l) were collected in Charleston, S.C. as well as in other laboratories of the Pasteurized Milk Network throughout the Southeast and U.S. The South Carolina readings rose by 57 percent, from 7 pCi/l in July 1970 to 11 pCi/l in July 1971. The corresponding change for the Southeast was a 10 percent rise, from 9 to 10 pCi/l, and for the U.S. as a whole, readings fell by 12 percent, from 8 to 7 pCi/l.

[44]A reading of 17 pCi/l was measured in the small dairies of North Augusta, South Carolina in June 1971 (July figures were not given) compared to a U.S. average reading of 9 pCi/l for June. See C. Ashley, "Environmental Monitoring at the Savannah River Plant," 1971 Annual Report (DPSPU 72-302), Table 5, p. 13 and *Radiological Health Data and Reports*, October 1971, Table 2, pp. 505–507.

[45]The probability is less than 2.5 percent that South Carolina's faster rise than the U.S. was a chance variation above the national norm. Infant mortality in South Carolina rose by 24 percent, from 24.6 deaths per 1,000 live births in January 1970 to 30.5 in January 1971. The corresponding change for the Southeast was a decline of 5 percent, from 26.3 to 24.9, and for the U.S. as a whole, a 3 percent decline from 21.7 to 21.1.

[46]The probability is less than 2.2 percent that South Carolina's slower decline in total deaths was due to chance. Total deaths in South Carolina fell by 2 percent from January 1970 to January 1971. Corresponding declines in the Southeast and the U.S. were 6 percent and 8 percent, respectively.

[47]The probability is less than 0.5 percent that South Carolina's faster rise than the U.S. was a chance variation above the national norm. South Carolina infant mortality rose by 15 percent, from 19.4 infant deaths per 1,000 live births in May–September 1970 to 22.3 in May–September 1971. The corresponding change for the Southeast was a slight decline from 21.43 to 21.37, and in the U.S. as a whole, there was a 3 percent decline, from 19.3 to 18.8.

[48]The probability is less than one in a million that the divergence of South Carolina's annual infant mortality from the 1968–73 U.S. trend was due to chance. After declining from 27.0 infant deaths per 1,000 live births in 1968 to 21.8 in 1971, infant mortality in South Carolina rose to 22.7 in 1973. Infant mortality fell steadily in the Southeast, from 24.9 in 1968 to 19.9 in 1973, and in the U.S. from 21.8 to 17.1.

[49]The probability is less than one in a million that this divergence was due to chance. South Carolina total mortality increased from 880 deaths per 100,000 people in 1968 to 910 in 1973. The corresponding change for the Southeast was from 960 to 970, and for the U.S. as a whole was from 970 to 940.

[50]In 1971, total mortality was 880 deaths per 100,000 people in South Carolina, 950 in the Southeast, and 930 in the U.S. as a whole.

[51]This estimated excess includes only those for which the probability is less than 1 percent that they were due to a chance variation, and adjusts for the age distribution of the population.

[52]Excess deaths as defined here include the number of observed deaths

from disease that are significantly higher than expected for each race-sex cohort compared to corresponding age-specific U.S. means (see the methodological appendix). Deaths in the estimate include only those for which there is less than 1 percent probability that they were due to chance. Unlike the earlier "total mortality" category, which includes deaths from "external" causes such as automobile accidents, acts of violence, and so forth, as well as deaths from disease, in the estimate of excess deaths, only mortality from disease is considered. This includes causes of death with International Classification of Diseases (ICDs) with a code of less than E800.

[53]The data are processed from the National Center for Health Statistics, *Mortality Surveillance System,* 1968–1983. The first 5-year period, 1968–73, excludes data for 1972, because they were only available as a 50 percent sample.

[54]Infant disease mortality was slightly, but not significantly, higher than the U.S. to start. Mortality from infant disease differs slightly from the standard "infant mortality" statistic used earlier. The rates of infant disease mortality excludes deaths from so-called "external" causes, such as automobile accidents, dropping, and other relatively infrequent non-disease causes of infant mortality.

[55]The category "birth defects" includes "congenital anomalies" and "perinatal complications" (ICDs 740–779).

[56]See Wilson B. Riggan, *et al., U.S. Cancer Mortality Rates and Trends 1950–1979,* Vol. 4, Washington, D.C.: U.S. Government Printing Office, 1987, p. 157.

[57]These are age-adjusted figures from *ibid.,* Vols. 1–3, averaged for the state groupings.

[58]North Augusta's June 1971 reading of 17 pCi/l of strontium-90 in milk rose to a 26 pCi/l maximum in December, and then fell to 9 pCi/l by the following December. See C. Ashley, op. cit., and C. Ashley and C. C. Zeigler, "Environmental Monitoring at the Savannah River Plant, 1972 Annual Report (DPSPU 73-302)," Table 5, p. 15.

[59]In September 1971, the average reading strontium-90 in public drinking supplies with surface water intakes within 25 miles of the plant was 6

pCi/l, which fell to one pCi/l the following October. See C. Ashley, *op. cit.*, Table 8 and C. Ashley and C. C. Zeigler, *op. cit.*, Table 8.

[60]These ratios are computed for picocuries of strontium-90 per gram calcium in each food type. See C. S. Klusek, "Strontium-90 in the U.S. diet, 1982 (EML-429)," New York, NY: USDOE, July 1984, Table 1 and Figure 6, pp. 9 and 19.

[61]Strontium-90 in the bone and flesh of Savannah River fish near the plant averaged at 10,000 pCi/kg in 1971. This equals 1,134 pCi per quarter pound of fish. The 1961 Federal Radiation Council maximum daily intake standard was 200 pCi/day. See *Radiation Data and Reports*, February 1974, p. 101.

[62]By ingesting 100 pCi, an infant receives a bone dose of 1.85 millirads. An intake of 1,134 pCi therefore represents a bone marrow dose of about 21 millirads. An infant receives a dose of about one millirad from a chest X-ray. A normal background dose is only 0.2 millirads per day. Nuclear Regulatory Commission, *Regulatory Guide 1.109: Calculation of Annual Doses to Man from Routine Releases of Reactor Effluents for the Purpose of Evaluating Compliance with 10 CFR Part 50*, Appendix 1 (NRC-NUREG 1.109), Washington, DC: NRC, Revision 1, October 1977, Table E14, p. 1.109–65.

[63]The average level of strontium-90 in the bone rose from 2.66 picocuries per gram calcium (pCi/g Ca) in 1970 to 3.85 pCi/g Ca in 1971 among children 10 years old and younger who were sampled in South Carolina. Corresponding readings in the Southeast fell from 2.92 to 2.59, and in the Northeast from 2.29 to 1.86. Readings of 4.10, 3.52, 6.93, and 3.78 pCi/g Ca in South Carolina were the highest recorded in the country for the 1st through 4th quarters of 1971. See *Radiological Health Data and Reports*, January, April, June and September, 1971 and *Radiation Data and Reports*, March, May and August, 1972 and January, 1973.

[64]See Gina Kolata, "New treatments may aid women who have repeated miscarriages," *The New York Times*, January 5, 1987.

[65]Ernest J. Sternglass, *Proceedings of the 6th Berkeley Symposium on Mathematical Statistics and Probability*, University of California Press, 1972, p. 145.

[66]See Jane E. Brody, "Natural chemicals now called major cause of disease," *The New York Times*, April 26, 1988 and Jean L. Marx, "Oxygen free radicals linked to many diseases," *Science*, Vol. 235, January 30, 1987, pp. 529–531.

[67]This division was derived from the analysis by Rita Fellers, "1979 Lung Cancer Mortality in South Carolina," Report prepared for Robert Alvarez, Environmental Policy Institute, July 1982.

[68]Only one measure was significantly higher in South Carolina than in the U.S.: the percentage of river mileage that failed to meet standards set for their use. But the higher-risk counties had better river quality (judged by this measure) than the lower-risk counties. Pesticide usage, on the other hand, was the only measure that was significantly worse in the higher-risk counties, but average usage in the state as a whole was less than in the U.S. Of the 34 pollution measures tested, 32 either had better readings or were insignificantly worse in the higher-risk counties vs. the lower-risk ones and in South Carolina vs. the U.S. The geographic comparisons were performed for estimates of toxic emissions to air, toxic discharges to water, toxic waste generation, closed toxic waste sites, as well as pesticides, air and water quality. Discriminant analysis and difference-of-means t-tests at the 95 percent confidence level were used for the geographic comparisons. For a detailed explanation of the methods and variables, see Public Data Access, Inc., *Toxics and Mortality Along the Mississippi River*, Washington, DC: Greenpeace USA, 1988, in which a similar multivariate analysis is performed of toxics contamination and health effects in counties along the River.

[69]See Tobacco Institute, *The Tax Burden on Tobacco: Historical Compilation*, Vol. 22, Washington, DC: Tobacco Institute, 1987, Table 11.

[70]The association of lung cancer with these activities was the result of well-known studies that began with W. J. Blot and J. F. Fraumeni, "Geographic patterns of lung cancer: industrial correlations," *American Journal of Epidemiology*, Vol. 103, 1976, pp. 539–550.

[71]See Riggan, *op. cit.*, pp. 154–155.

[72]See Kenneth J. Meier, *Regulation: Politics, Bureaucracy, and Economics*, New York, NY: St. Martins Press, 1985, p. 213.

[73]J. Tichler and C. Benkovitz, *Radioactive Materials Released From Nuclear Power Plants: Annual Report 1981*, Washington, DC: Nuclear Regulatory Commission, 1984.

[74]Robert S. Norris, Thomas Cochran and William Arkin, *Nuclear Weapons Data Book Working Papers: Known U.S. Nuclear Tests July 1945 to December 31, 1987*, (NWD-86-2 [Rev. 2b]), Washington, DC: Natural Resource Defense Council, September 1988, p. 40.

[75]Salt Lake City, Utah recorded 187 pCi/m^3, Boise, Idaho had 27 and the U.S. average was 1. *Radiological Health Data and Reports*, April 1971, p. 213.

[76]Percent changes are for July 1971 over July 1970.

[77]*Radiological Health Data and Reports*, April 1971, p. 206.

[78]E. J. Sternglass, *Secret Fallout: Low-level Radiation From Hiroshima to Three-Mile Island*, New York, NY: McGraw-Hill, 1981, p. 205.

[79]*Secret Fallout*, p. 221.

[80]*Secret Fallout*, p. 226.

[81]*Secret Fallout*, p. 228.

[82]In Pennsylvania, the infant mortality rate for the period January–March 1979 was 13.06 (infant deaths per 1,000 live births), which rose to 15.14 by the period April–July. Corresponding figures for Maryland were 8.66 and 12.23, for Upstate New York, 15.36 and 15.23, for New York City, 17.58 and 16.46, and for the U.S., 14.08 and 12.03. These data are from successive issues of *Monthly Vital Statistics Report* for the year 1979.

[83]The ratio of observed to expected mortality rates corresponds to 8.6 standard deviations.

[84]The average number of deaths during January–March 1979 in the tri-state area was 19,675, which rose to 20,083 in April–July.

[85]This estimated excess includes only those deaths for which the probability is less than 1 percent that they were due to a chance variation, and adjusts for the age distribution of the population. The expected trend was fitted to the log values of the observed mortality rates for the years 1970–1979.

[86]The increase in the U.S. crude mortality rate from 1979 to 1980 was 3.1 percent, compared to 5.5 percent in Pennsylvania, 5.2 percent for

New York, and 4 percent for the 21 states within 500 miles of TMI (corresponding to 5.8, 6.1, and 6.4 standard deviations respectively). Crude mortality rates are not adjusted for age.

[87]This excess share was calculated by subtracting the observed 1980 number of deaths in each state less the number that would have occurred had the 1980 crude mortality rate remained the same as in 1979.

[88]*Secret Fallout*, p. 258.

[89]Harvey Wasserman, "Three Mile Island did it: the fatal fallout from America's worst nuclear accident," *Harrowsmith*, May–June 1987.

[90]*Ibid.*

[91]*Ibid.*

[92]The 1979–80 infant mortality rate in Dauphin County was 18.2 deaths per 1,000 live births, compared to 13.3 in 1977–78; the corresponding numbers for the U.S. were 12.9 and 14.0, respectively.

[93]Infant birth defect mortality is defined here as the ratio of observed to expected infant deaths from congenital malformations. This is the same standardized mortality ratio (SMR) used in Chapter 4's discussion of South Carolina's birth defect mortality (see Table 4-1). The expected number of deaths are based on national age-race-sex-specific norms. In Dauphin County, the infant birth defect SMR increased from 1.03 during 1968–73 (not including 1972 because of an incomplete government sample) to 1.48 during 1979–83. In Pennsylvania, it increased from 1.06 to 1.12 during the same period. The significance of the difference between the change in Dauphin County's SMR and the change in Pennsylvania's SMR is tested with the following formula for the standard deviation of the difference between the two proportions: S.D. $= (p_o - p_e)$ / SQRT $[p_o(1-p_o)/N_o + p_e(1-p_e)/N_e]$. Where, p_o = observed proportion; p_e = expected proportion; SQRT = square root; N_o = observed number of deaths. This calculation yields a difference of 7.8 standard deviations.

[94]A. M. Hilton, editor, *Against Pollution and Hunger*, New York, NY: John Wiley and Sons, Inc., 1974, p. 127. According to the Centers for Disease Control, birth defects are becomingly an increasingly important cause of infant mortality, having increased by about twenty-three percent in 1986 over the rate during the 1970s. See *Morbidity and Mortality Weekly Report*, Vol. 38, No. 37, September 22, 1989.

[95]According to one reference: "Patients may complain of a metallic taste in the mouth within a few hours of the administration of a treatment dose of iodine-131, which is then excreted by their salivary glands." See Henry W. Wagner Jr., *Principles of Nuclear Medicine*, Philadelphia, PA: W. B. Saunders and Company, 1968, p. 359.

[96]The significance test for the difference between the change in the 10-county SMR and the change in Pennsylvania's SMR yielded the following results (in standard deviations): all diseases 2.5; heart diseases 6.1; infant diseases 8.6; birth defects 7.8; all cancers 7.8; lung cancers 4.9; breast cancers 12.4; child cancers 7.6.

[97]A multivariate examination of seventy-five alternative factors failed to identify a plausible rival hypothesis, including measures of toxic pollution, health services, socio-economic status, demographics, and others.

[98]E. I. du Pont de Nemours, "Savannah River Plant, January–December 1971," *Radiation Data and Reports*, February 1974, p. 102.

[99]*The New York Times*, October 4, 1988.

[100]*The Atomic Energy Act of 1946*, Public Law 585, 79th Cong., 60 Stat. 755–75; 42 U.S.C. 1801–1819, Sec. 10(b).

[101]Daniel S. Greenberg, *The Politics of Pure Science*, New York, NY: The World Publishing Co., 1969, p. 138.

[102]U.S. Department of Health, Education, and Welfare, *Radiological Health Data*, Vol. 4, No. 10, October 1963, p. 483. The publication was later renamed *Radiological Health Data and Reports* and then *Radiation Data and Reports*.

[103]For example, the January 1971 mortality data were first published in *Monthly Vital Statistics Report*, Vol. 20, No. 1 on April 2, 1971 whereas the revised data used in figures 4-3 and 4-4 were published in Vol. 21, No. 1 on March 28, 1972.

[104]In statistical terms, the South Carolina's original January 1971 infant mortality increase was 1.89 standard deviations higher than expected; when revised, it was 1.96 standard deviations. The original figure on total deaths was 1.84 standard deviations higher than expected; when revised, it was 2.02 standard deviations.

[105]In the final data, South Carolina infant mortality was 0.33 standard

deviations *less* than expected, and total deaths were only 0.08 standard deviations above expected.

[106]The number of infant deaths originally reported in *Monthly Vital Statistics Report* for South Carolina in January 1971 was 135. The revised figure (by place of occurrence) published a year later was also 135. The final number by place of residence was 97, 28% below the original and revised figures.

[107]The change in Pennsylvania's 1979 infant mortality rate from the period January–March to April–July diverged from the U.S. by 36 percent in the original by-occurrence data, but only by 10 percent in the final by-residence data. Similarly, the change in Pennsylvania's 1979 total deaths from the period January–March to April–July diverged from the U.S. by 1.2 percent in the original by-occurrence data, but only by 0.6 percent in the final by-residence data. Neither divergence was significant at the 95 percent confidence level, as reported in the final data.

[108]Montgomery, Alabama reported the following highest readings of gross beta radiation in precipitation (pCi/l) in the country during 1971: 186 in March, 75 in April, 148 in July, 28 in August, 108 in September, 98 in November, and 18 in December.

[109]See U.S. Environmental Protection Agency, *Environmental Radiation Data*, No. 10, Washington, DC: EPA, October 1977, ERAMS Section 1, p. 1.

[110]According to a *New York Times* report of May 23, 1987, by Michael Gordon, the Department of Energy admitted that there was an accidental release of fission products from the Mighty Oak underground test in April 1986, which was designed to test the ability of a laser device to beam radiation over long distances.

[111]Richard W. Clapp, S. Cobb, C. K. Chan, and B. Walker, Jr., "Leukemia near Massachusetts nuclear power plant," *The Lancet*, December 5, 1987, pp. 1324–1325.

[112]These releases have been associated with increases in low birth weights and infant mortality in nearby cities. See Ernest Sternglass, *Birth Weight and Infant Mortality Changes in Massachusetts Following Releases from the Pilgrim Nuclear Power Plant.* Submitted to the Massachusetts Legislature, June 10, 1986.

[113]A Freedom of Information Act request was submitted for the radiation data, asking for a computerized format. In response, EPA, which took over radiation monitoring responsibilities in the mid-1970s, stated that most of the data were never computerized, and that it would cost over $13,000 to comply with the request, even though the bulk of the data would be provided in paper copy. EPA Office of Radiation Programs, "Re: Freedom of Information Act Request RIN-6272-86," November 26, 1986.

[114]Lester B. Lave, S. Leinhardt and M. B. Kaye, "Low Level Radiation and U.S. Mortality, Working Paper 19-70-1," Pittsburgh, PA: Graduate School of Industrial Administration, Carnegie-Mellon University, July 1971.

[115]According to the final by-residence data in the *Vital Statistics of the United States*, the 1970 to 1971 change in South Carolina's May–September infant mortality rates diverged from that in the rest of the U.S. by 13 percent, compared to the 18 percent divergence recorded in the earlier revised by-occurrence data. Similarly, the divergence in total deaths for the same period fell from 4 percent in the revised data to 3 percent in the final data. Nevertheless, both divergences in the final data were significant at the 95 percent confidence level, which is to say that the probability that this divergence is due to chance is less than one out of twenty.

[116]Again, according to the final by-residence data in the *Vital Statistics of the United States*, the change in the tri-state area's 1979 infant mortality rate from the period January–March to April–July diverged from the rest of the U.S. by 8 percent, compared to the 36 percent divergence recorded in the original by-occurrence data. Similarly, the divergence in total deaths for the same period fell from 5.3 percent in the original data to 1.6 percent in the final data. Nevertheless, both divergences in the final data were significant at the 99 percent confidence level.

[117]Richard Rhodes, *The Making of the Atomic Bomb*, New York, NY: Simon and Schuster, 1986, p. 511.

[118]John Gofman, *An Irreverent, Illustrated View of Nuclear Power*, San Francisco, CA: Committee for Nuclear Responsibility, 1979, pp. 227–228.

[119]*The Washington Post*, April 14, 1979.

[120]Richard S. Norris, Thomas Cochran, William Arkin, *Known U.S. Nuclear Tests, July 1945 to December 1987*, Washington, DC: National

Resources Defense Council, 1988. The surprisingly high Soviet yield estimates are attributed to the French Ministry in Natural Resources Defense Council, *Nuclear Weapons Handbook, Vol IV*, New York, NY: Harper and Row, 1989, p. 373.

[121]Although mortality data prior to 1900 are less accurate, the secular decline exhibited here could be traced well back into the 19th century, with the discovery of antisepsis and the bacterial origin of infection. With extensions of longevity, the proportion of older persons increases. This aging causes the "crude" mortality rate, which is not adjusted for age, to flatten out. However, aging cannot explain the flattening out of infant and "age-adjusted" mortality rates observed during the period 1950 to 1965.

[122]Rachel Carson, *Silent Spring*, Boston, MA: Houghton Mifflin, 1962, p. 6.

[123]S. Postel, "Defusing the toxics threat: controlling pesticides and industrial waste," *Worldwatch Paper*, No. 79, September 1987, p. 8.

[124]Norris, *op. cit.*

[125]Andrei D. Sakharov, "Radioactive carbon from nuclear explosion and nonthreshold biological effects," *Soviet Journal of Atomic Energy*, Vol. 4, No. 6, June 1958.

[126]Abram Petkau, "Radiation carcinogenesis from a membrane perspective," *Acta Physiologica Scandinavia*, Supplement 1980, 492:81–90.

[127]T. Stokke, P. Oftedal, and A. Pappas, "Effects of small doses of strontium-90 on the ratbone marrow," *Acta Radiologica*, 1968, 7:321.

[128]Sakharov, *op. cit.*, p. 761.

[129]Linus Pauling, *No More War*, New York, NY: Dodd Mead, 1958.

[130]I. M. Moriyama, "The change in mortality trend in the U.S.," *Public Health Reports*, No. 1, Series 3, and I. M. Moriyama, "Recent changes in infant mortality trends," *Public Health Reports*, 1960, 75:391–406.

[131]In technical terms, the "R^2" of the asymptotic scenario indicates this curve explains about 96.7 percent of the observed variation, whereas the explanatory power of the parabolic scenario was slightly better, at 97.1 percent.

[132]S. Shapiro, E. R. Schlesinger and R. E. L. Nesbitt, *Infant, Perinatal, Maternal, and Childhood Mortality in the United States*, Cambridge, MA: Harvard University Press, 1968.

[133]K. S. Lee, N. Paneth, L. M. Gardner, M. A. Pearlman, "Neonatal mortality: analysis of recent improvement in the U.S.", *American Journal of Public Health*, Vol. 70, 1980, p. 17.

[134]*NCHS Bulletin*, 36, 17, Table 5, July 29, 1988.

[135]From 1983 to 1988, age adjusted mortality for all persons fell from 550.5 to 536.3 deaths per 100,000 people, for ages 15–24 it increased from 96.0 to 104.8, for ages 25–34 from 121.4 to 133.6, and for ages 35–44 from 201.9 to 217.6. See *NCHS Annual Summary 1988*, 37, 13, Table 5, July 26, 1989.

[136]C. S. Klusek, "Environmental Measurement Laboratory Report 435," New York, NY: USDOE, 1984.

[137]The ratios of change are derived by first dividing the mortality rate of each cohort at age 25–29 by that at age 5–9. This calculation yields 0.63 for all pre-1940 cohorts, meaning young-adult mortality was better than child mortality; for post-1940 cohorts, the calculation yields 1.70, meaning young-adult mortality was worse than child mortality. Second, the ratio for the later period is divided by that for the earlier one to derive a measure of change. The result is a ratio of change of 2.70 for all persons, which is shown, along with results for the race-sex groups, in the last row in table 7-1.

[138]M. Segi and M. Kurihara, "Cancer mortality for selected sites in 24 countries," *Japan Cancer Society*, Tohoku, Japan: Tohoku University, November 1972.

[139]F. Macfarlane Burnet, "Leukemia as a problem in preventive medicine," *New England Journal of Medicine*, Vol. 259, No. 9, pp. 423–431.

[140]J. P. Trowbridge and M. Walker, *The Yeast Syndrome*, New York, NY: Bantam Books, 1986.

[141]Lester B. Lave, S. Leinhardt, and M. B. Kaye, "Low-level radiation and U.S. mortality," Working Paper 19-70-1, Pittsburgh, PA: Carnegie-Mellon University Graduate School of Industrial Administration, July 1971.

[142]Ernest J. Sternglass, "Environmental radiation and human health," *Proceedings 6th Berkeley Symposium on Mathematical Statistics and Probability*, Berkeley, CA: University of California Press, 1972, pp. 145–216.

[143]Associated Press, April 2, 1987.

[144]Associated Press, April 3, 1987.

[145]Lynchburg, Virginia had readings of 180 picocuries per liter (pCi/1) of iodine-131 in rain, compared to 32 in Charleston, West Virginia, 24 in Wilmington, Delaware, and 29 and 19 in Middleton and Harrisburg, Pennsylvania.

[146]New York City's peak concentration was 32 pCi/1 and Philadelphia's was 22, compared with only 8 in Providence, Rhode Island, and 19 in Syracuse, New York. (These figures are for pasteurized milk; the figures reported in chapter 2 are for fresh milk.)

[147]Nuclear Regulatory Commission, "Safety evaluation and environmental impact appraisal by the office of nuclear reactor regulation, supporting amendment no. 29 to facility license no. DPR-44 and amendment no. 28 to facility license no. DPR-56, Philadelphia Electric Company Peach Bottom Atomic Power Station, units nos. 2 and 3, dockets nos. 50-2077 and 50-278," June 18, 1977, p. 4.

[148]A detailed chronology of Peach Bottom's problems has been compiled by the Susquehanna Valley Alliance and the Maryland Safe Energy Coalition.

[149]J. Tichler and K. Norden, *Radioactive Materials Released from Nuclear Plants: Annual Report 1983*, Washington, DC: Nuclear Regulatory Commission, 1986.

[150]In April 1987, total infant mortality was 38.6 deaths per 1,000 live births in Washington, D.C. and 11.1 in the U.S. In May 1987, total infant mortality was 24.5 deaths per 1,000 live births in Baltimore and 9.6 in the U.S.

[151]Infant mortality in Washington D.C. was reported to be only 9.3 deaths per live births in May 1987; the U.S. rate was 9.6.

[152]U.S. Department of Agriculture, *The Federal Milk Marketing Order Program*, Marketing Bulletin #27, June 1981.

[153]Telephone interview with Sidney Hall, Chief Food Protection Branch, Washington, D.C. Department of Consumer and Regulatory Affairs, March 8, 1989.

[154]Letter from Sidney Hall to Ben Goldman, March 9, 1989.

[155]Letter from Sidney Hall to Ben Goldman, March 15, 1989, and telephone interviews with dairy plants during April 1989.

¹⁵⁶U.S. Department of Agriculture, *Sources of Milk for Federal Order Markets by State and County,* January 1989, p. 4.

¹⁵⁷Consumption estimates use the national average per capita milk consumption rate of about 589 pounds annually.

¹⁵⁸In 1987, Pittsburgh's nonwhite infant mortality rate was 27.9 infant deaths per 1,000 live births, compared with the next highest rate of 23.7 for Detroit, according to the Pittsburgh Board of Health.

¹⁵⁹*Pittsburgh Post Gazette,* March 28, 29, and 30, 1989.

¹⁶⁰Gina Kolata, "New treatments may aid women who have repeated miscarriages," *The New York Times,* January 5, 1987.

¹⁶¹Jay M. Gould, B. Jacobs, C. Chen, and S. Cea, "Nuclear emissions take their toll," *CEP Newsletter,* December 1986.

¹⁶²The eight Northern-Midwestern states accounted for over half (55 percent) of the estimated national milk-radiation exposure risk. These states produced 41 percent of the nation's milk (circa 1982), and released 22 percent of the radiation from civilian reactors since 1974 (54.2 billion pounds of milk times 8.11 million curies of radiation yields an exposure risk of 440). Six Middle-Atlantic states (Delaware, Washington, D.C., Maryland, New Jersey, New York, Pennsylvania) followed with 39 percent of the national milk-radiation exposure risk, producing 17 percent of the nation's milk and 40 percent of radiation released (21.8 billion pounds of milk times 14.2 million curies yields an exposure risk of 309). Both regions' infant mortality improved significantly less since 1965–69 than the national trend. By contrast, infant mortality in the three Northern New England states (Maine, Vermont, New Hampshire) improved more rapidly than the nation as a whole, while they accounted for less than 3 percent of milk output and less than 1 percent of the radiation released (3.4 billion pounds of milk times 0.11 million curies yields an exposure risk of 0.38). Thus, the Northern-Midwestern states' risk of 440 is about 1,157 times greater than the Northern New England states' risk of 0.38. The milk production data are from: U.S. Bureau of the Census, *Statistical Abstract of the United States: 1986* (106th edition), Washington, DC: Government Printing Office, 1986.

¹⁶³The National Academy of Sciences has suggested that the 1958 Delaney Clause of the Food, Drug and Cosmetic Act, which bars the

addition to food of any substance known to cause cancer in humans or animals, would prohibit any trace of man-made radioactivity in food. This prohibition has never been enforced by responsible government agencies. See: Committee on Food Protection, *Radionuclides in Food*, Washington, DC: National Academy Press, 1973, pp. 78 and 86.

[164]J. Tichler and K. Norden, *Radioactive Materials Released from Nuclear Plants: Annual Report 1983*, Washington, DC: Nuclear Regulatory Commission, 1986.

[165]E. J. Sternglass, "Strontium-90 levels in the milk near Connecticut nuclear power plants," submitted to Congressman C. J. Dodd and Connecticut State Representative John Anderson, October 1977.

[166]Letter from Joseph M. Hendrie, Chairman of the Nuclear Regulatory Commission to Congressman Christopher Dodd, January 18, 1978.

[167]Cancer mortality in the four-county area rose by 30 percent from 1965–69 to 1975–82, compared to Connecticut's rise of 24 percent, and a U.S rise of 16 percent.

[168]*Hartford Courant*, March 14, 1987, p. 84.

[169]Lloyd Mueller, "An assessment of regional variation in Connecticut cancer rates, by proximity to nuclear power facilities," Hartford, CT: Connecticut Department of Health Services, 1987.

[170]*Hartford Courant*, July 17, 1987.

[171]Holger Hansen, *Connecticut Cancer Atlas*, University of Connecticut Health Center, 1988.

[172]Age-adjusted cancer incidence increased by 23.9 percent from 1963–72 to 1978–82 for males living in or near Middletown ("Area G") and by 18.7 percent for males living in or near Groton ("Area H"); both increases were significantly greater than the 10.9 percent rise for males in Connecticut as a whole. Age-adjusted cancer incidence increased by 3.1 percent during the same period for females near Middletown and by 17.7 percent for females near Groton; the latter increase was significantly greater than the 2.6 percent rise among Connecticut women as a whole.

[173]Dr. Hansen estimated that the rate of birth defects in Connecticut was 155.5 per 10,000 births in 1982–84, as against 116.2 in 1970–72. See Warren Froelich, "Rate of Down's Syndrome Increases Sharply in Connecticut," *Hartford Courant*, July 11, 1987.

[174]Letter to Dr. Gould from Charles Morgan, Connecticut Citizens Action Group, March 21, 1987.

[175]G. S. Habicht, G. Beck and J. L. Benarch, "Lyme Disease," *Scientific American*, July 1987.

[176]*U.S. Vital Statistics, 1985*, Vol. II, Part A, p. 8; *Health, United States, 1988*, March 1989, p. 53. Both published by the U.S. Public Health Service.

[177]George P. Georghiou of the University of California, Riverside compiled these figures, which were cited in S. Postel, *Defusing the Toxics Threat: Controlling Pesticides and Industrial Waste*, Washington, DC: Worldwatch Institute, 1987, p. 20.

[178]*Ibid.*, p. 19.

[179]E. J. Sternglass and J. Scheer, "Radiation exposure of bone marrow cells to strontium-90 during early development as a possible cofactor in the etiology of AIDS," Philadelphia, PA: American Association for the Advancement of Science, Annual Meeting, May 29th, 1986.

[180]Further support for this interaction is suggested by the surprising finding of French and German scientists in 1989 that "high ozone and acid-rain levels were found over African rain forest areas" that have no industrial emissions, but which do get heavy rainfall carrying fission products as well as industrial and other emissions. See *The New York Times*, June 19, 1989.

[181]Jean L. Marx, "The AIDS virus can take on new guises," *Science*, vol. 241, August 26, 1988, p. 1039–1040. In a report in the *New York Times* of June 9, 1988, Dr. Harold Schmeck attributed to Dr. Temple Smith the observation published in the June 9th issue of *Nature* that probably some 40 years ago "some genetic change turned the ancestral virus into the deadly AIDS virus of today." This is in full accordance with the Sternglass-Scheer hypothesis that the initial victims succumbing to the AIDS virus in 1980 were those persons born after 1945 whose immune systems may have been damaged by nuclear fallout.

[182]B. Schacter, M. M. Lederman, M. J. Levine, J. J. Ellner, "Ultraviolet radiation inhibits human natural killer cell activity and lymphocyte proliferation," *Journal of Immunology*, Vol. 130, No. 5, May 1983.

[183]J. A. Stoff and C. R. Pellegrino, *Chronic Fatigue Syndrome: The Hidden Epidemic*, New York, NY: Random House, 1988.

[184]Jane E. Brody, "Natural chemicals now called major cause of disease," *The New York Times*, April 26, 1988.

[185]*Secret Fallout*, p. 272.

[186]The Soviet Union ceased publication of its infant mortality rates for the years 1972 to 1975. However, the rate for 1976 was 31.1 infant deaths per 1000 live births, representing 38,700 more infant deaths than would have been expected with the 1971 rate of 22.9 deaths per 1000 live births. No official explanation for this ominous increase was offered. Yet during this period, Soviet reactors started up without the containment structures used in the U.S. to minimize radiation releases. The published Soviet data can be found in C. Davis and M. Fessbach, "Rising infant mortality in the U.S.S.R. in the 1970s," Series P-95, No. 74, Washington, DC: U.S. Census Bureau.

[187]*Investors Daily*, February 6, 1989.

[188]*New Scientist*, November 26, 1988.

[189]*The New York Times*, January 4, 1989.

[190]Marvin Resnikoff, *The Next Nuclear Gamble: Transportation and Storage of Nuclear Waste*, New York, NY: Council On Economic Priorities, 1983.

[191]*The New York Times*, January 8, 1989.

[192]Benjamin Friedman, *Day of Reckoning*, New York, NY: Random House, 1988, p. 85.

[193]*Secret Fallout*, p. 181.

[194]Bernard Rimland and Gerald Larsen, "Manpower quality decline: an ecological perspective," *Armed Forces and Society*, Fall 1981.

[195]Data on recent SAT scores, received from the College Board in New York City too late for the extended discussion in the text that they deserve, indicate that, as predicted by Sternglass, the average U.S. verbal score rose from an all-time low of 424 in 1980 to a peak of 431 in 1985 and 1986. This was matched by a corresponding increase from 466 to 475 in the average U.S. math score. But most alarming is a subsequent 4 point decline in the verbal SAT score to 427 in 1989, raising the question: what happened 18 years ago? The answer may lie in the fact that from October 1969 to October 1971, five underground tests in Nevada (Pod, Snubber, Mint Leaf, Baneberry and Diagonal Line) are known to have leaked at

least some 7 million curies of radiation into the atmosphere. The following states close to the Nevada Test Site displayed the following sharp declines in verbal SAT scores from 1985 to 1989: South Dakota -36; Wyoming -33; Montana -23; Arizona -21; and Oklahoma -21. The corresponding decline in SAT scores in far-off urban states was far more moderate: New York -8; New Jersey -2; Pennsylvania -6; District of Columbia -6. These wide variations in regional trends may help illuminate current concerns with the failure of higher education in America.

[196]C. Silverman, "Mental function following scalp X-irradiation for tinea capitatis in childhood," Washington, DC: Bureau of Radiation and Health of the U.S. Department of Health and Human Services, 1980.

[197]*Secret Fallout*, p. 195.

[198]R. J. Pellegrini, "Nuclear fallout and criminal violence: preliminary inquiry into a new biogenic predisposition hypothesis," *International Journal of Biosocial Research*, Vol. 9(21), pp. 125–143, 1987.

[199]"U.S. businesses brace for a disaster: a work force unqualified to work," *New York Times*, September 24, 25, 1989.

[200]Correspondence with Prof. Scheer in 1989 on his research underlying the publication in *The Lancet* of his findings on the effects of Chernobyl radiation on infant mortality in West Germany, as detailed in footnote 9.

[201]The average infant mortality rate for 1987 and 1988 was 5.8 deaths per 1,000 live births in Wyoming, 7.0 for Montana, and 10.0 for the U.S. See *NCHS Bulletin*, 37, 12, March 28, 1989.

[202]John Allen Paulos, *Innumeracy: Mathematical Illiteracy and its Consequences*, New York, NY: Hill and Wang, 1988, p. 105.

[203]For a more technical description of the methodology for calculating excess deaths, see Public Data Access, Inc., *Mortality and Toxics Along the Mississippi River*, Washington, DC: Greenpeace USA, 1988. The method used here is fully documented in the Greenpeace report with the one addition that race has been figured into the adjustment process.

[204]A. Petkau, "Effect of 22 Na^+ on a phospholipid membrane," *Health Physics*, Vol. 22, 1972, p. 239. See also A. Petkau, "A Radiation carcinogenesis from a membrane perspective," *Acta Physiologica Scandinavia*, Suppl. Vol. 492, 1980, pp. 81–90.

[205]A. Petkau and W. S. Chelack, "Radioprotective effect of superoxide

dismutase on model phospholipid membranes," *Biochemica et Biophysica Acta*, Vol. 433, 1976, pp. 445–456. See also A. Petkau, W. Kelly, W. S. Chelack, S. D. Pleskach, C. Barefoot, and B. E. Meeker, "Radioprotection of bone marrow stem cells by superoxide dismutase," *Biochemical and Biophysical Research Communications*, Vol. 67, No. 3, 1975, pp. 1167–1174; A. Petkau, W. S. Chelack and S. D. Pleskach, "Protection of post-irradiated mice by superoxide dismutase," *International Journal of Radiation Biology*, Vol. 29, No. 2, 1976, pp. 297–299; A. Petkau, "Radiation protection by superoxide dismutase," *Photochemistry and Photobiology*, Vol. 28, 1978, pp. 765–774; A. Petkau, "Protection and repair of irradiated membranes," in *Free Radicals, Aging, and Degenerative Diseases*, Alan R. Liss, Inc., 1986, pp. 481–508; and A. Petkau, "Role of superoxide dismutase in modification of radiation injury," *British Journal of Cancer*, Vol. 55, Suppl. VIII, 1987, pp. 87–95.

[206]Irwin Fridovich, "The biology of oxygen radicals: the superoxide radical is an agent of oxygen toxicity; superoxide dismutases provide an important defense," *Science*, Vol. 201, 1978, pp. 875–880.

[207]Charles Waldren, Laura Correll, Marguerite A. Sognier and Theodore T. Puck, "Measurement of low levels of X-ray mutagenesis in relation to human disease," *The Proceedings of the National Academy of Sciences*, Vol. 83, 1986, pp. 4839–4843.

[208]T. Stokke, P. Oftedal, and A. Pappas, "Effects of small doses of strontium-90 on the ratbone marrow," *Acta Radiologica*, Vol. 7, 1968, pp. 321–329.

[209]*Ibid.*

[210]Peter A. Cerutti, "Prooxidant states and tumor production," *Science*, Vol. 227, 1985, pp. 375–381.

[211]See New York Academy of Science, "Antioxidants may prevent or slow down heart disease," *Science Focus*, Vol. 3, No. 4, Spring 1989, p. 8, and Jane E. Brody, "Natural chemicals now called major cause of disease," *The New York Times*, April 26, 1988 and Jean L. Marx, "Oxygen free radicals linked to many diseases," *Science*, Vol. 235, 1987, pp. 529–531.

[212]Elizabeth Rosenthal, "Transplant patients illuminate link between cancer and immunity," *The New York Times*, December 5, 1989.

[213]*Ibid.*

[214]O. Heller and H. Wigzell, "Supression of natural killer cell activity with radioactive strontium: effector cells are marrow dependent," *Journal of Immunology*, Vol. 110, 1977, pp. 1503–1506. Also, E. Sternglass found in 1973 that significant changes in cervical cancer incidence and mortality in Baltimore women were directly correlated with changes in concentrations of the short-lived strontium-89 found in milk. See "Epidemiological studies of fallout and patterns of cancer" in *Radionuclides and Carcinogenesis, U.S. AEC Symposium Series 29, Conference-720505*, edited by C. L. Sanders, *et al.*, Washington, DC: U.S. Atomic Energy Commission, June 1973, pp. 254–277.

[215]Committee on the Biological Effects of Ionizing Radiation, *Health Effects of Exposure to Low Levels of Ionizing Radiation: BEIR V*, Washington, DC: National Academy Press, 1990.

[216]*Ibid.*, p. 218.

[217]*Ibid.*, p. 47.

[218]*Ibid.*

[219]*Ibid.*, p. 139 and Figures 3-4 and 3-5 on p. 146. Supporting references include: E. S. Copeland, editor, A National Institutes of Health Workshop Report, "Free radicals in promotion—a chemical pathology study section workshop," *Cancer Research*, Vol. 43, 1983, pp. 5631–5637; S. M. Fisher and L. M. Adams, "Suppression of tumor-promoter induced chemiluminescence in mouse epidermal cells by several inhibitors of arachinoic acid metabolism," *Cancer Research*, Vol. 45, 1985, pp. 3130–3136; B. O. Goldstein, G. Witz, M. Amoruso, D. S. Stone, and W. Troll, "Morphonuclear leukocyte superoxide anion radical (O^2) production by tumor promoters," *Cancer Letters*, Vol. 11, 1981, pp. 257–262; D. R. Jaffe, J. F. Williamson, G. T. Bowden, "Ionizing radiation enhances malignant progression of mouse skin tumors," *Carcinogenesis*, Vol. 8, 1987, pp. 1753–1755; J. B. Little and J. R. Williams, "Effects of ionizing radiation on mammalian cells," in S. R. Geiger, H. L. Falk, S. D. Murphy, and P. H. K. Lee, editors, *Handbook of Physiology*, Bethesda, MD: American Physiological Society, 1977, pp. 127–155; J. H. Marx, "Do tumor promoters affect DNA after all?" *Science*, Vol. 219, 1983, pp. 158–159; and J. E. Trosko, L. P. Yotti, S. T. Warren, G. Tsushimoto, and C. C. Chang, "Inhibition of cell-cell communication by tumor promoters," *Carcinogenesis*, Vol. 7, 1982, pp. 565–585.

[220]BEIR V, pp. 355–362. Supporting references include: W. J. Blot and R. W. Miller, "Mental retardation following in utero exposure to the atomic bombs of Hiroshima and Nagasaki," *Radiology*, Vol. 106, 1973, pp. 617–619, W. J. Blot, "Review of thirty years study of Hiroshima and Nagasaki atomic bomb survivors," II Biological effect. C. Growth and development following prenatal and children exposure to atomic radiation, *Journal of Radiation Research*, Vol. 16 (Suppl.), 1975, pp. 82–88, International Commission on Radiological Protection, *Developmental Effects of Irradiation on the Brain of the Embryo and Fetus: ICRP Publication 49*, Oxford: Pergamon, 1986, R. W. Miller and J. H. Mulvihill, "Small head size after atomic irradiation," *Teratology*, Vol. 14, 1976, pp. 335–338, M. Otaki and W. J. Schull, "In utero exposure to A-bomb radiation and mental retardation. A reassessment," *RERF Technical Report No. 1-83*, 1983, W. J. Schull and M. Otake, "Effects on intelligence of prenatal exposure to ionizing radiation," *RERF Technical Report 7-86*, 1986, United Nations Scientific Committee on the Effects of Ionizing Radiation (UNSCEAR), *Genetic and Somatic Effects of Ionizing Radiation: Report E. 86. IX. 9*, New York, NY: United Nations, 1986, and J. W. Wood, K. G. Johnson, Y. Omori, S. Kawamoto, and R. J. Keehn, "Mental retardation in children exposed in utero, Hiroshima and Nagasaki," *American Journal of Public Health*, Vol. 57, 1967, pp. 1381–1390.

[221]See W. J. Schull, M. Otake and Y. Yoshimaru "Effect on intelligence test score of prenatal exposure to ionizing radiation in Hiroshima and Nagasaki, A comparison of the old and new dosimetry systems," *1988 Revised RERF Technical Report 3-88*, In preparation.

[222]BEIR V, p. 362, reviewing: E. Ron, B. Modan, S. Flora, I. Harkedar, and R. Gureurt, "Mental function following scalp irradiation during childhood," *American Journal of Epidemiology*, Vol. 116, 1982, pp. 149–60.

[223]BEIR V, p. 8.

[224]See footnote 163.

[225]BEIR V, p. 376, reviewing: V. E. Archer, "Association of nuclear fallout with leukemia in the United States," *Archive of Environmental Health*, Vol. 42, 1987, pp. 263–271.

[226]BEIR V, p. 387, reviewing: E. G. Knox, A. M. Stewart, E. A. Gilman,

and G. W. Kneale, "Background radiation and childhood cancer," *Journal of Radiology Protection,* Vol. 8, No. 1, 1988, pp. 9–18.

[227]BEIR V, p. 6.

[228]One from the U.S. is: J. K. Lyon, M. R. Klauber, J. W. Gardner, and K. S. Udall "Childhood leukemias associated with fallout from nuclear testing," *New England Journal of Medicine,* Vol. 300, 1979, pp. 397–402.

[229]M. J. Gardner and P. D. Winter, "Mortality in Cumberland during 1959–78 with reference to cancer in young people around Windscale (letter)," *The Lancet,* Vol. i, 1984, pp. 216–217, and M. J. Gardner, A. J. Hall, S. Downes, and J. D. Terrell, "Follow up study of children born to mothers resident in Seascale, West Cumbria (birth cohort)," *British Medical Journal,* Vol. 295, 1987, pp. 822–827.

[230]E. Roman, V. Beral, L. Carpenter, *et al.,* "Childhood leukemia in the West Berkshire and Basingstoke and North Hampshire District Health Authorities in relation to nuclear establishments in the vicinity," *British Medical Journal,* Vol. 294, 1987, pp. 597–602, and D. J. Hole and C. R. Gillis, "Childhood leukemia in the west of Scotland," *The Lancet,* Vol. 2, 1986, pp. 525.

[231]BEIR V, p. 379, reviewing: S. Openshaw, M. Charlton, A. W. Craft, and J. M. Birch, "Investigation of leukemia clusters by use of a geographical analysis machine," *The Lancet,* Vol. i, 1988, pp. 272–273.

[232]J. L. Lyon, and K. L. Schuman, "Radioactive fallout and cancer (letter)," *Journal of the American Medical Association,* Vol. 252, No. 14, 1984, pp. 1845–1855, C. J. Johnson, "Cancer incidence in an area of radioactive fallout downwind from the Nevada test site," *Journal of the American Medical Association,* Vol. 251, 1984, pp. 230–236, G. G. Caldwell, D. Kelley, M. Zack, H. Falk, and C. W. Heath, "Leukemia among participants in military maneuvers at a nuclear bomb test: a preliminary report," *Journal of the American Medical Association,* Vol. 244, 1980, pp. 1575–1578, G. Caldwell, D. Kelley, C. W. Heath Jr., and M. Zack "Mortality and cancer frequency among military nuclear test (Smoky) participants, 1957 through 1979," *Journal of the American Medical Association,* Vol. 250, No. 5, 1983, pp. 620–624, G. Caldwell, D. Kelley, C. W. Heath, Jr., and M. Zack, "Polcythemia vera among participants of a nuclear weapons test," *Journal of the American Medical Association,* Vol.

252, 1984, pp. 662–664, and S. C. Darby, G. M. Kendall, T. P. Fell, *et al.*, "A summary of mortality and incidence of cancer in men from the United Kingdom who participated in the United Kingdom's atmospheric nuclear weapon tests and experimental programs," *British Medical Journal*, Vol. 296, 1988, pp. 332–338.

[233]BEIR V, p. 378, reviewing: D. Forman, P. Cook-Mozaffari, S. Darby, *et al.*, "Cancer near nuclear installations," *Nature*, Vol. 329, 1987, pp. 499–505 and P. Cook-Mozaffari, F. L. Ashwood. T. Vincent, *et al.*, "Cancer incidence and mortality in the vicinity of nuclear installations in England and Wales, 1950–1980," *Studies on Medical and Population Subjects, No. 51*, London: Her Majesty's Stationery Office, 1987. An earlier study had not found a clear pattern of cancer increases in individuals living near fourteen nuclear facilities and five non-nuclear plants in England and Wales. See J. A. Baron, "Cancer mortality in small areas around nuclear facilities in England and Wales," *British Journal of Cancer*, Vol. 50, 1984, pp. 815–829.

[234]See footnote 111.

FIGURE 2-1 source: EPA. *Environmental Radiation Data Report.*

FIGURE 2-2 derived from: NCHS, *Monthly Vital Statistics Report.*

FIGURE 2-3 derived from: NCHS, *Monthly Vital Statistics Report.*

FIGURE 2-4 derived from: EPA, *Environmental Radiation Data Report* and NCHS, *Monthly Vital Statistics Report.*

FIGURE 2-5 derived from: EPA, *Environmental Radiation Data Report*, NCHS, *Monthly Vital Statistics Report*, and data submitted to the authors by Jens Scheer.

FIGURE 2-6 derived from: Data submitted to the authors by Jens Scheer.

FIGURE 2-7 source: David F. DeSante and Geoffrey R. Geupel, "Landbird productivity in central coastal California: the relationship to annual rainfall and a reproductive failure in 1986." *The Condor*, Vol. 89, 1987, p. 641.

FIGURE 4-1 derived from: EPA, *Radiological Health Data and Reports.*

TABLE 4-1 derived from: NCHS, *Mortality Surveillance System.*

FIGURE 4-2 derived from: EPA, *Radiological Health Data and Reports* and EPA, *Radiation Data and Reports.*

TABLE 4-2 derived from: C. Ashley, "Environmental Monitoring at the Savannah River Plant, 1971 Annual Report (DPSPU 72-302)," Tables 4, 5, and 8, pp. 11–15, C. S. Klusek, "Stontium-90 in the U.S. Diet, 1982

(EML-429)," New York: U.S. Department of Energy Environmental Measurements Laboratory, July 1984, Table 1, p. 9, and U.S. EPA, *Radiation Data and Reports*, February 1974, Tables 12, 14, and 15, pp. 99–101.

FIGURE 4-3 derived from: NCHS, *Monthly Vital Statistics Report*.

FIGURE 4-4 derived from: NCHS, *Monthly Vital Statistics Report*.

FIGURE 4-5 derived from: NCHS, *Vital Statistics of the United States*.

FIGURE 4-6 derived from: NCHS, *Vital Statistics of the United States*.

FIGURE 4-7 derived from: NCHS, Mortality Surveillance System.

FIGURE 5-1 derived from: "Assessment of offsite radiation doses for the Three Mile Island unit accident, TDR-TMI-116," prepared for the Metropolitan Edison Company by Pickard, Lowe and Garrick, Washington, DC: July 31, 1979.

TABLE 5-1 derived from: NCHS, *Mortality Surveillance System*.

FIGURE 5-2 derived from: NCHS, *Vital Statistics of the United States*.

FIGURE 5-3 derived from: NCHS, *Vital Statistics of the United States*.

FIGURE 5-4 derived from: NCHS, *Vital Statistics of the United States*.

FIGURE 5-5 derived from: NCHS, *Mortality Surveillance System*.

FIGURE 6-1 derived from: NCHS, *Monthly Vital Statistics Report* and NCHS, *Vital Statistics of the United States*.

TABLE 6-1 source: NCHS, *Monthly Vital Statistics Report*.

FIGURE 6-2 derived from: NCHS, *Monthly Vital Statistics Report* and NCHS, *Vital Statistics of the United States*.

TABLE 6-2 source: NCHS, *Monthly Vital Statistics Report*.

FIGURE 6-3 derived from: NCHS, *Monthly Vital Statistics Report*.

FIGURE 6-4 derived from: NCHS, *Monthly Vital Statistics Report*.

FIGURE 6-5 derived from: EPA, *Radiological Health Data and Reports* and EPA, *Radiation Data and Reports*.

FIGURE 6-6 derived from: EPA, *Environmental Radiation Data Report*.

FIGURE 6-7 derived from: EPA, *Environmental Radiation Data Report*.

FIGURE 7-1 derived from: NCHS, *Vital Statistics of the United States*.

TABLE 7-1 derived from: NCHS, *Vital Statistics of the United States*.

FIGURE 7-2 derived from: NCHS, *Vital Statistics of the United States*.

FIGURE 7-3 derived from: NCHS, *Vital Statistics of the United States*.

FIGURE 7-4 derived from: NCHS, *Vital Statistics of the United States.*
FIGURE 7-5 derived from: NCHS, *Vital Statistics of the United States.*
FIGURE 7-6 derived from: NCHS, *Vital Statistics of the United States.*
FIGURE 8-1 derived from: EPA, *Environmental Radiation Data Report.*

FIGURE 8-2 derived from: NCHS, *Vital Statistics of the United States* and data submitted to the authors by the Maryland Department of Human Health and Hygiene.

FIGURE 8-3 derived from: NCHS, *Vital Statistics of the United States* and J. Tichler and K. Norden, *Radioactive Materials Released from Nuclear Plants: Annual Report, 1983*, Washington, DC: Nuclear Regulatory Commission, 1986.

FIGURE 8-4 derived from: NCHS, *Monthly Vital Statistics Report* and data submitted to the authors by the Maryland Department of Human Health and Hygiene.

FIGURE 8-5 source: Federal Milk Market Administrator, "Federal Order No. 4: Annual Statistical Report 1987," Alexandria, VA.

FIGURE 8-6 derived from: NCHS, *Vital Statistics of the United States* and Census Bureau, *Statistical Abstract of the United States.*

Non-fiction from **Four Walls Eight Windows**

Bachmann, Steve.
Preach Liberty: Selections from the Bible for Progressives. pb: $10.95.

David, Kati.
A Child's War: World War II Through the Eyes of Children. cl: $17.95.

Dubuffet, Jean.
Asphyxiating Culture and Other Writings. cl: $17.95.

Gould, Jay, and Goldman, Benjamin.
Deadly Deceit: Low-Level Radiation, High-Level Cover-Up. cl: $19.95.

Hoffman, Abbie.
**The Best of Abbie Hoffman: Selections from
"Revolution for the Hell of It," "Woodstock Nation,"
"Steal this Book," and Other New Writings.** cl: $21.95.

Johnson, Phyllis, and Martin, David, eds.
Frontline Southern Africa: Destructive Engagement. cl: $23.95, pb: $14.95.

Jones, E.P.
Where Is Home? Living Through Foster Care. cl: $17.95.

Wasserman, Harvey.
Harvey Wasserman's History of the United States. pb: $8.95.

To order, send check or money order to Four Walls Eight Windows, PO Box 548, Village Station, N.Y., N.Y. 10014. Add $2.50 postage and handling for the first book and 50¢ for each additional book. Or call 1-800-835-2246, ext. 123.